Environmental Risk Communication

Principles and Practices for Industry

SUSAN ZUMMO FORNEY
ANTHONY J. SADAR

CRC Press
Taylor & Francis Group
Boca Raton London New York

CRC Press is an imprint of the
Taylor & Francis Group, an **informa** business

Second edition published 2021
by CRC Press
6000 Broken Sound Parkway NW, Suite 300, Boca Raton, FL 33487-2742
and by CRC Press
2 Park Square, Milton Park, Abingdon, Oxon, OX14 4RN

First edition published by CRC Press 1999

CRC Press is an imprint of Taylor & Francis Group, LLC

ISBN: 978-0-367-46977-1 (hbk)
ISBN: 978-1-003-08344-3 (ebk)

Typeset in Palatino LT Std
by KnowledgeWorks Global Ltd.

To the three guys who own my heart and make everything worthwhile: Miles (thank you for taking on more than your fair share to make this possible), Quinton, and Connor. And to my beloved family members who have always been a source of strength and comedic relief.

SZF

To my four sweethearts (see first edition) and two grandsons, Killian and Thomas.

AJS

Contents

List of Figures

List of Tables

Preface to Second Edition

Environmental Risk Communication: Principles and Practices for Industry first hit the shelves in 1999 in the wake of the Clean Air Act Risk Management Plan (RMP) rules. Professionals from a wide variety of industries and operations, many of which had previously enjoyed relative obscurity in the public domain, were suddenly faced with communicating to stakeholders about the frightening consequences of chemical accidents. The first edition offered straight-forward, principled guidance for these professionals that was incorporated into risk communication and media training programs.

Since the first days of RMP, the field of environmental risk communication has continued to mature, as have the public participation mandates of environmental regulations in general. The current public involvement landscape features citizen science, power-wielding non-governmental organizations, and viral social media.

On its surface, environmental risk communication has sprouted new varieties of challenges and opportunities. Underneath, however, the principles that govern sound practice remain largely neglected. Projects that undertake public participation in name only are reworked, delayed, canceled, or litigated to conclusion at great expense. Companies that spend money on simply polishing reputations remain untrusted and challenged. Opportunities for mutual benefit between facilities and their host communities are lost in unproductive battles.

This second edition refreshes the conversation on principled stakeholder communications and extends recommended approaches and practices beyond the RMP to all environmental programs managed by today's industries. Its aim is straight-forward, no-frills guidance geared to industry owners and operators; environmental, health and safety professionals; and industry technical and communication staff. On top of candid advice and examples from more than three decades of our own experience in environmental risk communication, we have condensed the best and most relevant guidance and data from practitioners and researchers from across the globe. It is at once novel and conventional, and it is everything we would want a client to be equipped with before engaging in environmental risk communication.

Acknowledgments

The authors would like to acknowledge Mark D. Shull for his contributions as co-author of the first edition of *Environmental Risk Communication:Principles and Practices for Industry,* and Peter Sandman (www.petersandman.com) for graciously and freely sharing the collection of his works with anyone wishing to learn more about risk communication.

About the Authors

Susan Zummo Forney is founder of EHS Communication Partners, Inc., where she provides expert communications counsel, quality training programs, and professional communication products to clients dealing with environmental, health and safety concerns. She has more than 30 years of experience in environmental consulting, with an emphasis in communications and compliance. Her approach provides a unique blend of technical knowledge and principled communication strategies that position clients for success in stakeholder communications and engagement.

Ms. Zummo Forney has provided risk communication and community relations support for industries such as electric utilities, chemical manufacturers, and waste management companies, as well as for programs under the U.S. Environmental Protection Agency and Department of Defense. She has conducted numerous communications presentations and training workshops for industry trade groups, professional organizations, and regulatory agencies, and has been published in the *Journal of Environmental Health* and *EM* Magazine.

Ms. Zummo Forney is a member of the International Association of Public Participation and the Air & Waste Management Association (A&WMA) Allegheny Mountain Section, for which she pens monthly newsletter columns on communication and water-quality compliance.

She holds an M.S. in Environmental Science and Management from Duquesne University and a B.S. in Science with a Certificate of Minor in Science, Technology, and Society from Penn State University.

Anthony J. Sadar, founder of Environmental Science Communication, LLC, is a Certified Consulting Meteorologist who has advised on numerous high-profile and controversial projects involving communications on air quality and emergency release planning. He recently retired from serving as an Air Pollution Administrator for the Allegheny County Health Department, Air Quality Program, in Pittsburgh, PA. He is currently Adjunct Associate Professor at Geneva College in Beaver Falls, PA. Mr. Sadar's 40 years of experience includes work as a manager of air projects for a scientific research firm and as a science and technology instructor for several colleges and universities.

Mr. Sadar has presented air monitoring and modeling results before general public audiences, some quite hostile. He learned from theory and practice some of the best techniques to address highly-charged issues and has trained industry and agency staff to successfully manage challenging public meetings and other risk communication efforts.

Mr. Sadar has published many technical articles and has contributed to several classroom manuals. He is a member of the Local Emergency Planning Committee of Allegheny County, the American Meteorological Society, and the A&WMA, where he was on the Editorial Advisory Committee of *EM* magazine, A&WMA's monthly publication for environmental managers.

Mr. Sadar holds an M.S. in Environmental Science from the University of Cincinnati, an M.Ed. in Education from the University of Pittsburgh, and a B.S. in Meteorology from Penn State University.

Mr. Sadar was the principal author of the first edition of *Environmental Risk Communication: Principles and Practices for Industry* (CRC Press/Lewis Publishers, 2000), in which he collaborated with veteran communication practitioner Mark D. Shull.

Introduction

When it comes to communicating about environmental risk, all organizations, large or small, must find their footing somewhere between the options of keeping their heads in the sand and hiring public relations firms to handle communications for them. Why somewhere in between? Because, outside of simple luck, neither of these approaches usually ends well.

While there are many useful tools for helping with your communication endeavors, there is no replacement for starting with a solid foundation. Industries that wish to continue operating in host communities particularly have much to lose—more than money, that is—from investing in quick fixes.

As demonstrated throughout this book, communication is more than just talk. The best way to improve perceptions about your industrial operations, counteract negative publicity campaigns, and foster constructive community relations is to create and follow a principled plan of action. The principles proffered in the first edition of *Environmental Risk Communication* focused largely on orienting and assisting facilities with the communication challenges associated with the Clean Air Act Risk Management Program. However, they continue to apply broadly today. They are:

- Operate your facility legally and ethically.
- Educate employees on the operation's benefits and risks.
- Listen carefully and respond appropriately to public concerns.
- Be open, honest, considerate, but cautious with the news media.

This second edition dives deep into the critical but often neglected foundational elements of sound environmental risk communication. The first half of this book addresses four key recommendations that help build strong foundations in stakeholder communication and engagement:

1. Walk the talk.
2. Set goals.
3. Know your audience.
4. Deal with emotional elements first.

Walking the talk means acting with integrity, letting your actions speak for you, being authentic, paying homage to the clean-hands doctrine, and even knowing when to stand firm against political pressure. Without this, everything else becomes window dressing.

Setting goals means that every investment of time and money is geared toward achieving your definition of success. Not every organization will have the same goals, and thus cookie-cutter communication plans provide little value.

Understanding your audience is a much-talked about but little practiced concept. Communicating without knowing your audience is likely to result in your message missing its mark. Moreover, it can waste valuable time and resources, as not every stakeholder should carry equal weight when it comes to your communication investment.

Finally, communicating about technical issues without dealing with the emotional components of risk perception is like rolling a boulder uphill. Many of the strategies for dealing with emotional roadblocks feel counterintuitive at first, which is why you need to invest time in understanding them.

The second half of this book offers guidance on the process and tools of the trade, as well as for dealing with the tough cases and working with the mass media. The concluding chapter offers best practices for conveying information that can be referred back to over time for different circumstances, including preparing for crises.

Case studies and lessons learned are weaved throughout this book to offer examples of these principles and practices at work. Written during the COVID-19 pandemic, this book also includes some every-day examples from this nationwide challenge in risk-communication that touched every citizen across the nation.

This book is intended to be both a grounding in enduring principles and a continued resource for best approaches and techniques. In a field of practice with no magic bullets and no guarantees, your best bet is being principled, informed, prepared, and flexible.

1

The Art and Science of Risk Communication

Introduction

The ability to communicate effectively is a prized attribute in the work force and in society at large. Even without the high stress and controversy often associated with communicating about environmental risk, challenges abound in transmitting and receiving information. This chapter briefly addresses the broader challenges of communication and then examines the definition and approach to environmental risk communication that will be taken throughout the remainder of this book.

1.1 Challenges of Communication

While communicating about risk requires some special considerations, the practice shares three main attributes with all forms of communication, in that it is:

- *Imprecise*—We transmit ideas, concepts, and facts imperfectly the first time based on our own biases or filters. And then every time this information is translated by another individual or group, the message is further filtered and eroded.
- *Irreversible*—Historically, this meant once you say or print something, you cannot take it back. Today it carries a new meaning with social media. This is true today more than ever. Everything and anything is memorialized in an instant with social media, available any time for unearthing and sharing with the world.
- *Contextual*—We frame information based on our experience and point of view. The framing occurs both when we transmit information and when we process it as receivers.

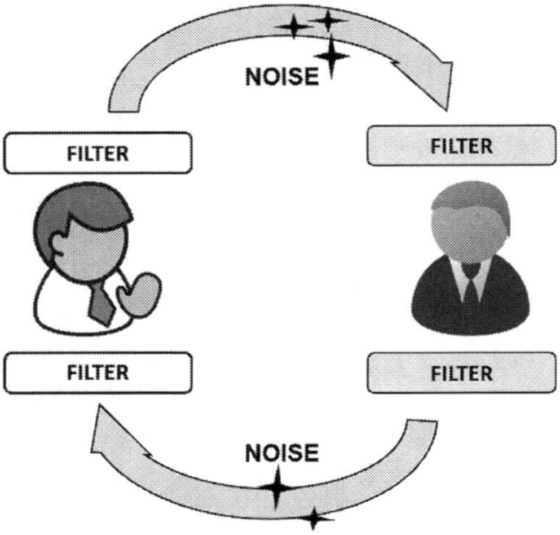

FIGURE 1.1 Communication is imprecise.

As the simple illustration in Figure 1.1 shows, you cannot assume that the information you transmit is being received exactly as intended. Both the senders and the receivers experience interference not only from outside influences but also from internal biases that cause them to shape what they say and hear. Among these factors are simple influences such as background noise, current state of mind, tiredness, stress, age, and so on.

Have you ever played the telephone game where one person whispers a message into the ear of another, and then that person whispers the message to another, and so on, until you get to the last person in line? The message uttered by the final receiver is often quite different from the one shared by the initial sender. This potential for faulty transmission is why people employed in the nuclear industry use three-way communication. When they issue instructions, they ask the receiver to repeat the instructions back, and then they confirm if they are correct or not. This protocol must be followed whether you are uttering two words or multiple sentences—there is simply too much at stake to get it wrong.

While we readily acknowledge that framing and interference occurs, we continue to manage our everyday communications as if everything gets through exactly as we planned. And we are often taken by surprise when something we thought was crystal clear was interpreted differently by others.

In *Why Don't People Listen*, Hugh Mackay (2013/1994) dives more deeply into how and why we may miss the mark when communicating with others. Among what he refers to as the *ten laws of human communication* addressed in his book is the tendency of listeners to interpret messages in terms of their own perspective of the world and their own circumstances; in this way they affirm what already exists in their belief system and discount what does not.

He also discusses how listeners who feel insecure in their relationship with the speaker are not likely to listen well.

Mackay further addresses the importance of perceived balance in the relationship between speakers and listeners. When listeners' attitudes are challenged head-on, they are likely to defend and reinforce those attitudes. In addition, if they are not consulted first about a change that affects them, they are more likely to resist the change for that simple fact.

As we unfold the principles and practices of environmental risk communication throughout this book, you will find evidence of these laws in many of our recommendations.

1.2 Risk Communication Defined

The EPA's website (www.epa.gov/risk/risk-communication) describes risk communication in this way:

> Risk communication is the process of informing people about potential hazards to their person, property, or community. Scholars define risk communication as a science-based approach for communicating effectively in situations of high stress, high concern or controversy.
>
> From the risk manager's perspective, the purpose of risk communication is to help residents of affected communities understand the processes of risk assessment and management, to form scientifically valid perceptions of the likely hazards, and to participate in making decisions about how risk should be managed. Risk communication tools are written, verbal, or visual statements containing information about risk.

The situations of *high stress, high concern,* and *controversy* included in the EPA's definition are often what comes to mind when one thinks of communicating about environmental risk issues. Indeed, these are the components that typically raise sufficient concern for individuals and organizations to seek professional assistance.

Also notable in the EPA's definition are the words *hazard* and *risk.* These terms are often confused or used to mean the same thing. However, when you are communicating about risk, it is helpful to distinguish between them. Hazard is the potential of something to cause harm (sickness, injury, death, damage, etc.). Risk is the mere probability or chance that a particular hazard will cause that harm. For example, concentrated sulfuric acid is a hazardous chemical. However, if it is handled appropriately, the risks it poses are small.

When engineers, scientists, and executives ran into difficulties communicating about hazards and risks in the late 1970s, the prevailing thought about the source of the problem was that the public was incapable of understanding complex technical information. So, early communication efforts were focused

Four basic risk communication challenges (Covello & Allen, 1988):

1. Challenges with the information communicated.
2. Challenges with the source of information.
3. Challenges with the channels of communications.
4. Challenges with the receivers of information.

Interestingly, the challenges of *communicating* about environmental risks bear great resemblance to the technical model for evaluating risk called the Source-Pathway-Receptor (SPR) model, which describes the flow of an environmental pollutant from a source, through different pathways (or channels) to potential receptors (or receivers).

largely on attempting to simplify information and convince members of the public that risks were low. However, the factors that create challenges for risk communication, that is, high stress, high concern and controversy, have little to do with technical knowledge or complexity. In actuality, people generally trust technical professionals to get the science and technology right. It is the emotional dimensions that cause most of the problems.

The art and science of risk communication has grown out of the idea that, instead of just assuming the public is uninformed, there is a logic to how the public perceives risk, and that perception is shaped by a number of factors unrelated to hazard and risk. These factors include:

- Complex issues and conflicting science.
- Lack of trust in the source(s) of information.
- Poor track records or legacy issues.
- Public misperception and fear.
- Emotions that overrule facts.
- Influence from the media and the Internet.
- Personal biases.
- Competing agendas.

1.3 Categories of Risk Communication

Depending on your circumstances, your reason for communicating about risk may differ. For instance, you may be dealing with members of the public who are upset about the real or perceived threats from an operation you are planning. Or, you may be working to get out emergency evacuation orders related to an accidental release. Whatever the situation, there are four general categories of risk communication within which your effort may lie:

- *Precaution/Care advocacy*—This type of risk communication applies to situations where your organization has a vested interest in elevating the concerns of stakeholders about a risk so that they are more likely to take a desired action. For instance, you may be working to improve employee adherence to personal protective equipment rules.
- *Communication upkeep*—This category of risk communication applies to situations where no special concerns exist on either the part of the organization or its stakeholders. Under these circumstances, organizations should be continuing efforts to build trust and relationships with stakeholders. They should also be using existing communication channels to monitor for opportunities and potential challenges that may be brewing.
- *Outrage management*—This type of risk communication describes situations that often come to mind when talking about risk communication, where stakeholders are upset. As discussed in Chapter 5, *Address Emotions Before Facts*, the key to dealing with these situations is actually to put technical discussions temporarily on hold.
- *Crisis management*—This category of risk communication describes situations where organizations are communicating about risk in the midst of accidents and disasters. The pressure on organizations and stakeholders in these situations changes the manner in which information should be transmitted, as well as audiences' ability to process the information.

All four types of risk communication are addressed throughout the remainder of this book in the context of the principles and practices discussed in Chapters 2–9.

1.4 Orientation to Principled Risk Communication

When risk communication was first discussed as a practice more than 40 years ago, much of the attention was focused on *fixing* the audience. The question "How do we get an unsophisticated public to understand this complex information?" was being asked over and over again in conference rooms everywhere.

As the practice of risk communication matured, it went through a number of stages. First came the realization that the public would be more trusting if people had direct access to information rather than simply the conclusions drawn by experts. Next came the understanding that public stakeholders needed some context for the results of risk assessments. Following that came the stage where organizations realized the importance of listening to public stakeholders and treating them as a legitimate participant in the process.

Despite all the progress made in effective risk communication techniques, much of the focus of organizations remains on fixing the audience—priming them to be more receptive of organizations' predetermined positions on environmental matters and projects. This approach not only glosses over the true intent of stakeholder communications and public participation, it also results in missed opportunities for mutually beneficial outcomes.

Based on our own experience, as well as that of the many well-respected authors and practitioners cited in this book, the greatest gains you can make in the realm of communications and relationship building come from looking inward first. Stephen Covey, who is often quoted in this book, once said "As long as you think the problem is out there, that very thought is the problem."

Covey (1989) points out in *The 7 Habits of Highly Effective People* that people may temporarily get by with superficial behaviors and commitments, but eventually true motives surface and relationships fail. For industries operating in host communities and wishing to establish solid brands and reputations, thinking long-term and big-picture is a must. "What we *are*," says Covey, "communicates far more eloquently than anything we *say* or *do*." Put another way in a quote attributed to Ralph Waldo Emerson, "What you are shouts so loudly in my ears I cannot hear what you say."

Thus, the first and primary principle discussed in this book is walking the talk and operating with integrity. When you operate from a point of integrity, it becomes easier to set clear goals, align expectations, meet regulatory obligations, talk and act with consistency, and keep commitments—all things crucial to building trust and credibility.

With the proper foundation in place for communicating about risk, the techniques themselves become merely best practices rather than a means of manipulating your stakeholders. They are proven methods that help you communicate more effectively, assist you in dealing with roadblocks, and prevent you from inadvertently sabotaging your efforts. The better grounded an organization is, the easier it can navigate communication landscapes where one-size-fits-all solutions do not apply.

References

Covello, V. and Allen, F., *Seven Cardinal Rules of Risk Communication, U.S. Environmental Protection Agency*, Office of Policy Analysis, Washington, D.C., 1988.

Covey, S. 1989. *The 7 Habits of Highly Effective People*. Rev. Ed. New York, NY: Simon & Schuster Fireside.

Mackay, H. 2013; 1994. *Why Don't People Listen?* Sydney, Australia: Pan Macmillan.

2

Walk the Talk

Introduction

In the face of stiff economic competition and ever-increasing public demands for transparency, staying true to core principles is not only good business acumen, it is a commodity for enduring operations.

Companies that operate safely, legally, ethically, and with integrity but that do not communicate well regarding environmental and health risks are at a disadvantage. But companies that do not operate this way yet spend thousands on sleek risk communication efforts can be wasting their time and money, particularly in the long run.

From not building houses on sand to practicing what you preach and walking the talk, idioms abound for constructing strong core principles and sticking to them. Sound communication programs should start with nothing less.

Compliance with all environmental and safety regulations must lie at the lowermost level of a company's operating commitments. Otherwise, operators come to the court of public opinion with unclean hands. Given the complex regulatory landscape of environmental, health, and safety regulations, even the most well-intentioned companies can fall short of full compliance and proper accident prevention. Thus, taking steps to ensure that your facilities are as compliant and prepared as you think they are is a good investment.

Above all, organizations desiring to fare well in the court of public opinion must operate with integrity. When core values and beliefs align at all levels of your organization, they can be communicated, exemplified, and enforced. They serve as the backbone for all decision making, including those involving environmental stakeholders.

2.1 Comply With All Regulations

While organizations are well aware of their obligation to comply with regulations, doing the hard work of dotting all the "*i*"s and crossing all the "*t*"s is another matter.

Owners and operators face a myriad of complex and often confusing business and environmental statutes and regulations.

Typical business regulations include:

- Wage and Hour laws and regulations.
- Fair Labor Standards Act.
- Fair Employment Law.
- Americans with Disabilities Act.
- Occupational Safety and Health Administration (OSHA) regulations.
- Fire Protection and Life Safety Code requirements.

Typical environmental regulations include those addressing:

- Waste management and disposal.
- Air emissions.
- Wastewater and storm water discharges.
- Water encroachment and obstruction.
- Hazardous substances.
- Public right-to-know reporting.
- Accidental release planning and control.
- Dam safety.
- Land use and natural resource protection.
- Local zoning.
- Dangerous cargo and hazardous material transportation.
- Toxic substances control.
- Endangered species protection.
- Environmental Justice.

Rules and regulations affecting business and industry are promulgated at the federal, state, county, and municipal levels. While most regulations are readily available to the public over the Internet, identifying every requirement that applies to your company can be a real challenge, as can connecting the dots between the various regulatory authorities.

For example, consider regulations promulgated under the Clean Air Act (CAA) (as amended November 1990). State and local air-pollution agencies must follow suit with regulations that are at least as stringent as the federal regulations. These state and local authorities develop state implementation plans (SIPs) outlining how they will carry out the intent of the Act. SIPs must be evaluated and approved by the federal government. As an example of the interrelationship and complexity of government agencies,

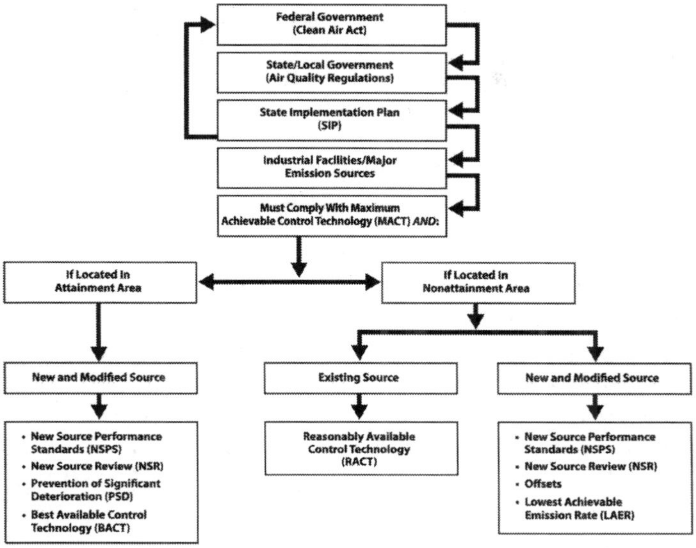

FIGURE 2.1 Air agency and industry relationships. (Adapted from Management side of engineering, Plant Eng., 49(4), 104, 1995. With permission.)

their responsibilities, and the responsibility of the regulated community, an overview of the relationship between air agency programs and industrial facilities that emit major quantities of air pollutants is provided in Figure 2.1 (adapted from Sadar, 1995).

SCOPE OF RESPONSIBILITIES WITHIN REGULATORY AGENCIES

In implementing laws, federal, state and/or local agencies carry out a number of responsibilities in engaging with regulated entities. Using the Clean Air Act as an example, the span of government responsibilities includes:

- Plans to achieve compliance (State Implementation Plan).
- Development and oversight of emission inventories.
- Modeling of source impacts.
- Development of emission standards.
- Monitoring of ambient air quality.
- Approval of compliance schedules.
- Issuance of construction and operating permits.

- Collection of fees.
- Investigation of complaints.
- Inspections.
- Review of operating schedules, practices, and records.
- Enforcement of regulations.
- Imposition of fines and penalties.

The responsibilities of regulatory authorities vary depending on which agency has primacy; sometimes state or local authorities may be responsible for implementing one or more parts of a law while EPA maintains primacy in other parts. In many cases, agency personnel who perform permitting functions are separate from those working in reporting or compliance and enforcement. Thus, companies often deal with multiple individuals both within and between agencies.

2.1.1 Considering the Cost of Compliance

Not only is the myriad of compliance obligations daunting, the costs can be enormous. These may include:

- Permit and approval fees.
- Engineering, technical, and legal consulting fees.
- Other professional service fees (for example, laboratory analyses and equipment maintenance).
- Expenditures on site preparation and pollution control equipment.
- Personnel time and expenses (including new projects and ongoing compliance assurance).
- Training costs.
- Personal protective gear.

As a result, many companies postpone compliance costs as long as possible. Unfortunately, this approach becomes more and more expensive for the following reasons:

- Agencies are facing continuous budget cuts and thus are increasingly using enforcement penalties as a means of generating revenue.
- Agencies have less flexibility in granting compliance schedules because of increased public scrutiny and third-party lawsuits.
- Consulting and legal fees balloon under rush situations, as more resources are thrown in to meet deadlines.

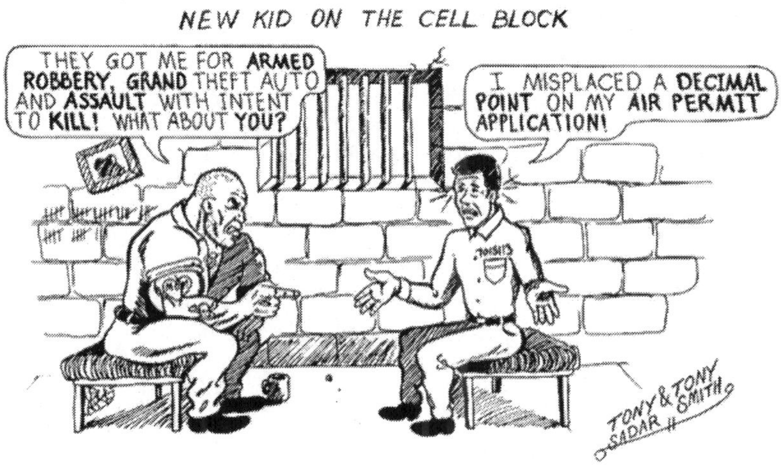

Postponing compliance obligations often leads to cascading violations as companies try to play catch-up and force solutions into existing circumstances. Where compliance lags, costs build up from penalties associated with violations and enforcement actions and fees from consultants and attorneys. Where willful or repeat violations exist, costs may also come in the form of civil and criminal convictions. Outside these obvious costs, there are a number of less tangible damages associated with non-compliance. These include:

- Diminished public image.
- Loss in employee morale.
- Weakened position for proposing future projects that are potentially controversial.

2.1.2 Benefits of an Integrated Compliance Approach

As with most anything in life, an ounce of prevention can be worth a pound of cure. Proactive environmental compliance programs share some common characteristics:

- Environmental policy that ties to a company's core mission.
- Clear organization with personnel responsibilities/accountabilities at all levels of the organization.
- Record keeping and documentation program that includes plans, reports, and training.
- A full accounting or audit of regulatory requirements that apply to the facility's operations. For processes or equipment that fall into

potentially gray areas, regulatory determination letters or documented expert evaluations are kept on file.

- Pollution prevention and waste minimization policies that are integrated into production and compliance planning. (This may include recycle/reuse/reduce, material substitution/reformulation, and waste exchange initiatives.)
- Well documented operational processes to allow for knowledge sharing and transfer among personnel.
- Emissions/Discharge inventories that are based on actual operating conditions (or at least as much defensible data that may be obtained to validate models).
- Means of investigating and correcting incidents and noncompliance.
- Means of continuing program evaluation and improvement.
- Means of monitoring emerging issues.
- Public involvement and community outreach.

As noted in the list, integration is a key element of an effective environmental management system. The less integrated that environmental planning and compliance programs are with the rest of the organization, the less opportunity that exists for recognizing chances to improve performance and cut environmental compliance costs over the long run.

In *Making Business Sense of Environmental Compliance*, Singh (2000) presented the results of four case studies in which companies lost millions of dollars because they managed environmental compliance as a function separate from other facility planning functions. As shown in Figure 2.2, there are several drawbacks of managing environmental compliance as an isolated

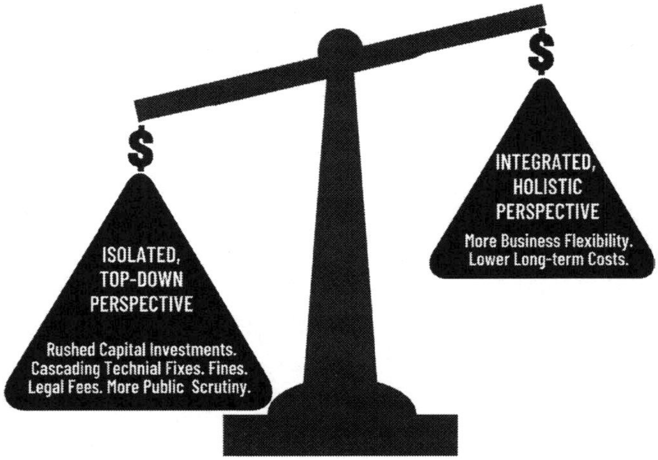

FIGURE 2.2 An integrated environmental compliance approach costs less in the long run.

corporate function responsible for implementing top-down business mandates. These drawbacks include the high cost of rushed capital investments and the escalation of non-compliance as environmental engineers seek technological band-aids to meet new pollution standards. Says Singh,

> A truly integrated approach takes an entire-plant perspective. It reviews new regulations, plant operations, emerging technologies, changes in markets, and fluctuations in product demand *simultaneously* to find solutions that best address both environmental compliance and production concerns...In this holistic approach, the environment becomes a criterion for making business decisions, and business needs become criteria for making environmental decisions.

Singh's evaluation found that without the pressure of enforcement actions, companies typically do not even think about evaluating the profitability of processes or products. When faced with major regulations, they usually leave technical solutions up to environmental engineers, that is, until insurmountable costs force consideration of operational changes.

The following recommendations are based on a review of the cases examined by Singh:

- Where possible, eliminate marginal products or processes that generate a disproportional amount of pollution.
- View emerging regulations with an eye toward possible opportunities to eliminate waste and/or make production more efficient.
- Seek coordinated input from marketing, operations, and environmental staff when planning to meet future product and service demands.
- When facing major regulatory changes, consider whether changing production technology makes sense. Sometimes in the long run, the cost to update technology is cheaper than retrofitting it with additional treatment systems. (According to Singh, "The best time to install new technology is when many factors have the potential to converge, such as low productivity, availability of more efficient technologies, and forthcoming environmental regulations.")
- Consider the consequences of generating and managing different types of waste before pursuing new business. This is particularly true for industries that have centralized waste or water treatment facilities.
- Calculate the real, or equitable, cost of product lines or waste streams because they could contribute more than their share of pollution or other problems. Without knowing the true costs of production, companies can miss opportunities for reducing pollution, negotiating higher prices with customers, cutting costs, or ceasing unprofitable activity.

MODELING THE COST OF NONCOMPLIANCE

Estimating the financial gains from postponing expenditures requires a model similar to the EPA's BEN model, which calculates the economic benefit of noncompliance (Singh, 2000). While this model may not be valid for many manufacturing situations, it may be worth evaluating to help design site-specific cost-benefit analyses for long-range planning. The model may be downloaded for free at https://www.epa.gov/enforcement/penalty-and-financial-models.

As regulatory requirements are ever evolving, monitoring of emerging issues, both at the federal and more local levels, is an important element of an integrated management system. Opportunities to provide comment and respond to early information requests should be pursued wherever possible. In addition, companies should take advantage of opportunities to be on advisory committees formed by government agencies where members help to inform and shape regulatory initiatives.

Many federal and state bureaus offer online resources and email alerts for regulatory updates and proposed legislation. One particularly helpful resource is *Inside EPA*, which provides a series of newsletters that keep readers informed of the news and insight pertinent to EPA activities across programs.

At the local level, both compliance and community relations efforts require boots on the ground. Someone from your facility should be attending municipal meetings and zoning board hearings. Not only does this familiarize you with the local project approval process, it allows you to build relationships with local officials and residents, as well as to be available should people have questions or concerns about your facility. (These are the most fruitful exchanges because they have the potential to nip in the bud legitimate complaints as well as to correct misinformation early.) Companies that factor in the concerns and interests of local officials and neighbors (where these align with company business objectives) have a great advantage over those who are unaware of these interests.

2.2 Operate With Integrity

This book stresses the theme of operating with integrity. Without it your communication efforts will eventually falter. Thus, we focus on principles and practices that lead not only to effective communication but also to establishing relationships with your stakeholders, those right outside your fence line and far beyond.

As will be discussed later in this book, decisions involving environmental risk are inextricably tied to values and ethics. In a compendium from the 1994 meeting of the American Chemical Society entitled *Environmental Risk Decision Making: Values, Perceptions and Ethics*, C. Richard Cothern offered this on the connection:

> Values and ethics should be included in the environmental decision-making process for three reasons: they are already a major component, although unacknowledged; ignoring them causes almost unsurmountable difficulties in risk communication; and because it is the right thing to do.
>
> Values and value judgements pervade the process of risk assessment, risk management, and risk communication as major factors in environmental risk decision making. Almost every step in any assessment involves values and value judgements. However, it is seldom acknowledged that they even play a role....
>
> Cothern, 2019/1996

These important points are driven home further in the book's Preface with pertinent observations and recommendations by retired research chemist Paul A. Rebers, the co-organizer of the Environmental Risk Decision Making conference. Rebers concludes:

> If we can recognize that values, ethics, and perceptions, as well as scientific data enter into the process of environmental risk decision making, we will have made an important step forward. This should make it easier for the public to understand how difficult and indeterminate the process may be. It should also make them demand to know the biases as well as the expertise of those making decisions. By being completely honest with the media and the public, we are making an important step in gaining their confidence...
>
> Cothern, Editor, 2019/1996

Regarding the expertise and communication of decision makers and gaining public confidence, crisis communications and public relations veterans Alan Bernstein and Cindy Rakowitz note that:

> If you are prepared to face the worst-case scenario and own up to the truth without qualification or excuses, your constituents will respect you as an authority. Anything less places you at risk that your communications audience will find ways to tear you apart.
>
> Bernstein & Rakowitz, 2012

No amount of risk communications consultation or training can take the place of personal integrity and ethical behavior—that must come from each of us as individuals. Consider your values, value judgments, ethics, and perceptions, in general and specifically regarding the operations in your charge.

Chances are that they match those of your stakeholders. For instance: "Would you have a problem if your family was located next to the plant? Would you be concerned if this operation was outside your back window?" These sorts of questions come up frequently in one form or another at public meetings, and you should be ready to answer them honestly, *and hopefully negatively.*

TRUST IS EARNED...

"...through sustained, observable behavior, not by engaging in promiscuous displays of self-congratulation about how much a company cares."

Glass Jaw, Dezenhall, 2014

When you exemplify, communicate, and enforce strong core values in your organization, you position your employees to serve as good ambassadors and you also strengthen the company's ability to weather the storm in times of crisis. In *Emergency Public Relations* by Bernstein and Rakowitz, the chapter titled "Core Values," reveals how "strong core values reinforce a brand's ongoing positive perception even when something goes wrong." The common successful formula for "the companies and individuals whose images survived their blunders" is:

> They immediately addressed their situations with an apology, a deep compassion, and a recitation of their strong core values. They made clear that their core values ran deeper than their current difficulties, placing them in the category of a temporary lapse in their normally good judgment. Indeed, each stressed that they had developed these laudable core values long before they committed the current round of blunders.

The authors go on to observe:

> Core values prevent problems. They mitigate the impact of problems on reputations and image. They define who we are and how we respond. Core values help reduce liabilities after negative events. They facilitate the necessary return to normalcy. They force organizations to own their own actions and own up to them.
>
> Bernstein & Rakowitz, 2012

Do not forget that core values follow you home at night, go with you to local shops and the golf course, and accompany you on business trips. Walk the talk in front of everyone, everywhere, at all times. Colleagues and staff take cues from you. Equally as important, employees can become disgruntled due to circumstances beyond your control; thus, what you once thought was between them and you can become very public very quickly.

2.3 Be Authentic

Perceptive people can spot a phony. Insincere individuals, although they may be charismatic, should not be a part of your communications team. In Chapter 3 of *Winning Your Audience,* Rosebush (2020) stresses the importance of "discovering your authentic self."

Rosebush asserts that "the essential element that leads to acceptance by any audience [is] authenticity." He continues with identifying the key to the development and adoption of authenticity:

> Self-knowledge is the key. It will keep you from speaking on a topic about which you know too little, using a vocal tone or phrasing in a way that is pretentious or not true to your natural voice—or even standing in a way that throws off or intimidates your audience.
>
> Rosebush, 2020

In other words, find out who you are and be yourself.

GENUINE AUTHENTICITY

"Your genuine authenticity is what will ultimately win your audience."

Winning Your Audience, Rosebush, 2020

In striving to be authentic, do not become overly concerned about how your true self will measure up when communicating with the public. Not everyone will be suited for speaking at public meetings, handling on-camera interviews, or facilitating advisory panel meetings. (The skill sets required for those roles are addressed in Chapters 8 and 9.) Should your organization lack in-house talent for a particular need, you can always reach out to a qualified consultant for direct assistance or training.

2.4 Above All Else...Be Humble

As is often said, we saved the best for last. This chapter began by stressing the importance of integrity. Germane to integrity is humility.

The importance of humility is emphasized in a recent best-selling management book, *Leadership Strategy and Tactics: Field Manual* (Willink, 2020). This book was one of the most popular books ever sold on Amazon.com, which is quite a feat. The last section of Willink's book proffers reasonable advice on communications (see the *Frontline Communication Guidance* insert).

FRONTLINE COMMUNICATION GUIDANCE

- Keep stakeholders informed.
- Address rumors with good information.
- Frame messages with the perspective of the listener in mind.
- Deliver the truth tactfully.
- Check inappropriate emotions but speak with poignancy.
- Explain actions and apologize if things go wrong. Tell what is being done to correct and prevent future missteps. Do not cover up mistakes.
- Be approachable but careful with words.
- Remain humble.

Adapted from Leadership Strategy and Tactics: Field Manual, Willink, 2020.

Willink asserts that "there is one type of person who can never become a good leader: a person who lacks humility." And, Willink reminds leaders (and we can add communicators) of the reality that they "might be above their subordinates in the rank structure, but they are not *actually superior* to those below them in the chain of command, and this means leaders must respect them."

The history of an exemplary person drives this point home. Ms. Katherine Johnson, the National Aeronautics and Space Administration gifted mathematician and one of the subjects of the 2016 movie "Hidden Figures," died at the age of 101 in February 2020. In a *Wall Street Journal* obituary on her, it was noted that when she was young, her father told her: "You are as good as anybody…but you're no better" (Hagerty, 2020).

Acting with humility will serve both you and your stakeholders well. It opens your mind to new ideas and helps others set aside the defenses that often get in the way of genuine dialogue.

CASE IN POINT

No Polish Needed

In discussing community relations services with a consulting firm, a junior environmental engineer with a small manufacturing company was impressed with the consultant's expertise, range of services, and print samples. His company, he explained, was rather unsophisticated in this department. As an example, he said, about a month ago the plant had experienced a rare upset that resulted in a minor but odorous air emission release. The plant manager had promptly gone door-to-door to let people know what happened in case they smelled the odor and became concerned.

This engineer may not have realized it at the time, but the plant manager's gut instinct, quick response, and forthright communication with neighbors was the best prescription any risk communication consultant could have offered.

References

Bernstein, A. B., and C. Rakowitz 2012. *Emergency Public Relations: Crisis Management in a 3.0 World*. Fourth Edition. Bloomington, IN: Xlibris Corporation.

Cothern, C. R., Editor. 2019/1996. *Handbook for Environmental Risk Decisions: Values, Perceptions, & Ethics*. Boca Raton, FL: CRC Press/Lewis Publishers.

Dezenhall, E. 2014. *Glass Jaw: A Manifesto for Defending Fragile Reputations in an Age of Instant Scandal*. Rev. Ed. New York, NY: Twelve.

Hagerty, J. R. 2020. Human Computer Found Niche at NASA, Obituaries: Katherine Johnson, 1918–2020, *The Wall Street Journal*. February 29, p. A12.

Rosebush, J. 2020. *Winning Your Audience: Deliver a Message With the Confidence of a President*. Rev.Ed. New York, NY: Center Street.

Sadar, A. J. 1995. "Demand to Curb Pollution is Getting Everyone in on the Act–the Clean Air Act." *Plant Engineering* 49 (4): 104.

Singh, J. 2000. Making Business Sense of Environmental Compliance. *MITSloan Management Review* (Spring 2000). https://sloanreview.mit.edu/article/making-business-sense-of-environmental-compliance/ (accessed July 31, 2020).

Willink, J. 2020. *Leadership Strategy and Tactics: Field Manual*. Rev. Ed. New York, NY: St. Martin's Press.

3

Set Goals to Guide Your Communication Investment

Introduction

From marketing to plant operations to compliance and safety, every corporate department, every important initiative, has a written plan with goals. Risk communication efforts, unfortunately, often fall through the cracks somewhere between marketing and environmental, health and safety departments. They bubble up from permit applications, accidental releases, emerging contaminant concerns or similar issues. They often suffer from lack of sound strategies, commitment, and resources. Ironically, many end up consuming vast resources, as attorneys and consultants are hustled on board for damage control.

While risk communication efforts typically stem from episodic needs, they should be handled in a proactive, not reactive, manner. Proactive planning helps you target resources, build trust, and support the long-term interests of your organization. Reactive risk communication, on the other hand, often wastes time and money, erodes trust, and jeopardizes progress.

3.1 Align Your Plans to Your Mission

Operating with integrity entails consistently communicating and modeling core values at every level within the organization. Thus, all communication plans should reflect organizational missions and visions.

As an example, nearly all corporate visions identify operating in a sustainable manner as a core principle. This might manifest in the angle or selection of stories written for social media feeds. As another example, commitment to transparency might result in recommending earlier or more public information sessions than what's required by law when permitting a new operation.

CORPORATE MISSION LANGUAGE EXAMPLES

"Our vision is to be the world's leading coatings company by consistently delivering high-quality, innovative and sustainable solutions that customers trust to protect and beautify their products and surroundings. This vision will guide us on our journey toward our common goals and principles."

PPG Vision, http://corporate.ppg.com/our-company.aspx

"...being a good corporate citizen is not only the right thing to do, it is also essential to our success." "If we are to succeed in business, we must do it on principles that are honest, fair, lawful and just."

Principles espoused in 2017 U.S. Steel Sustainability Report, https://www.ussteel.com

Being a good neighbor might mean that a higher percentage of charitable dollars be earmarked for local giving or hands-on projects rather than for regional or national donation campaigns. It may also mean that you invite your local fire department and first responders to practice mock emergencies with you. In these examples, walking the talk is both incorporated into planning and on display for stakeholders to see. As shown in Table 3.1, aligned, proactive communication programs reap numerous benefits over piecemeal, reactive programs.

TABLE 3.1

Proactive Versus Reactive Communication

Proactive Communication	Reactive Communication
Thinks long-term	Thinks short-term
Builds good will for future dialogue	Jeopardizes future good will for short-term gains
Sets the stage for success	Ignores tripping hazards and obstacles
Achieves buy-in and consistency at all organizational levels	Lacks buy-in and consistency across organization
Targets resources where they are most effective	Scatters resources among seemingly pressing issues
Builds collective knowledge of successful strategies and actions	Fades from organizational memory

3.2 Know Your Purpose and Objectives

No pen should touch paper, no finger should strike a key, until you know what you want to achieve. As a rule of thumb, you should include three, and no more than five, goals for your communication plans. Each goal should be:

- *Aligned*—consistent with the company's mission statements and policies.
- *Actionable*—doable and recordable, not lofty and esoteric.
- *Appropriate*—meets your needs and supports your objectives.
- *Realistic*—obtainable under your circumstances and restrictions.
- *Measurable*—can be quantified and tracked.

Most communications practitioners will advise that you break your guiding information into goals, objectives, and strategies. Goals describe where you want to wind up. Objectives are specific steps you take to achieve your goals, and strategies are the road maps for your objectives. For most projects, however, combining the goals, objectives and strategies into "purpose and objectives" or "strategy and approach" will sufficiently guide your plan.

By putting your goals in writing, you provide common ground for everyone involved in the communication effort, including consultants. Everyone should be on the same page, and should understand what a *win* would look like for your effort.

As an example, a goal in a transmission line siting project might be to *achieve a public participation process that exceeds both regulatory requirements and the stated expectations of the siting board.* A supporting objective might be to *make every effort with stakeholders to find an acceptable route such that a reasonable third-party observer would conclude that the stakeholders had every opportunity to find an acceptable route.* In this case, a supporting strategy might be to *conduct joint fact finding with stakeholders in selecting route alternatives.*

Approaches to communication efforts will vary depending on the purpose and the audience you want to reach. These in turn will directly impact the messages you develop. Similar to purpose and objectives, key messages will provide lane markers and guardrails for your efforts (more on developing key messages in Chapter 4).

Once you have established your purpose and objectives, you can determine the actions required to support them (often referred to as "tactics"). This is your to-do list for executing your plan. While many risk communication efforts start with a to-do list, the most effective begin with purpose.

3.3 Set Timelines and Budgets

No plan is complete without a timeline and a budget. For corporate communication plans and crisis communication plans, a timeline may simply be a requirement to review and update the plan once per year. In this case, the budget would cover the review plus any annual costs. For example, annual costs for your crisis communication plan may include communications training and/or desktop or mock drills. These may or may not be tied to emergency drills conducted by your environmental, health, and safety (EHS) department.

For project-specific risk communication efforts, it may be tempting to simply run an open-ended campaign on the assumption it will conclude with the project. However, binding your project with a budget and timeframe provides more clarity and objectivity. If things are going fine, you can always extend or rerun the plan, with or without adjustments.

Most importantly, written timelines and budgets secure advanced approval by upper management. Without the approval of those controlling purse strings, communication efforts are usually last in line for funding and first in line at the chopping block—despite the fact that business and risk managers consistently rate communication as a core competency and poor communications as a major business risk (Project Management Institute, 2013). Moreover, without being a line item in an annual budget, proactive communication plans go by the wayside, which means even crisis communication plans gather dust and fade from memory.

3.4 Plan to Evaluate and Realign

Communication efforts take time and resources. They are rarely one-offs, and they can impact more than your bottom line. Thus, they are worth being evaluated—both to realign current efforts where necessary and to make future efforts more successful.

One obvious means of evaluation is to determine whether or not you met your objectives. Another means of evaluation is to revisit the research you conducted at the outset and compare the metrics. For instance, if you relied on news clippings and interviews to gauge stakeholder sentiment before starting, you can do so again to measure where you stand at the conclusion. Unless a communication plan covers a rapid time period, it is often helpful to conduct a midcourse evaluation to determine whether you are on target. A midcourse evaluation allows you to realign your efforts and reassign resources to get the biggest bang for your buck. A note of caution here: While adjusting tactics makes sense, changing your metrics does not. If metrics are changed midcourse, post-project evaluations lose their value.

3.5 What is In a Risk Communication Plan

Communications planning at the corporate level is often tied to marketing and public relations initiatives. Thus, plans at this level may include SWOT analyses (strengths, weaknesses, opportunities, threats) and other marketing evaluations not directly related to environmental and health risk communication. Other elements of corporate communication plans, however, mirror components of facility and project plans—purpose, background, roles and responsibilities, strategies, action items, schedule, budget, and evaluation. Table 3.2 lists information typically contained in a project communication plan involving environmental and health risks.

3.6 Planning for a Crisis—It's Not If, But When

Work at your facility is going along normally. Then, without warning, something terribly wrong happens. The normal flow of business comes to an abrupt halt. Without a crisis communication plan, chaos may ensue, and what may have turned out to be a minor emergency turns into a prolonged crisis.

As Steven Fink (2013) notes in *Crisis Communications: The Definitive Guide to Managing the Message,* crisis in business is as inevitable as death and taxes; it is not a question of *if,* but rather *when.* Crises come in all shapes and sizes, from major

TABLE 3.2

Example of Information to Include in a Risk Communication Plan

Purpose and Objectives
Stakeholders
Communication Team
Communication Risks and Mitigations
Key Messages
Communication Tools and Activities
Action Timeline and Calendar
Estimated Costs
Performance Evaluation
Attachments/Inserts/Stand Alones:
Project team contacts
Media contacts
Elected officials
Mailing list
Website and social media policies
Project fact sheets
News releases and standby statements

accidental releases to rumor mongering by a disgruntled employee to natural disasters. Handling the facts on the ground can be a challenge. With the added pressures of social media, cell phone videos, and outrage culture, it can be a calamity.

> Never being caught up in a controversy may not mean that you are doing everything right, it could just be that you have been lucky so far.

Fortunately, however, you have strategies and tools at your disposal to help prepare your business for navigating a crisis. Use them, and you can minimize the damages, possibly even coming out stronger on the other side. Ignore them, and you increase your chances of a scorched reputation, a bruised bottom line, or worse.

"RACE" MODEL–BUILD IT *BEFORE* YOU NEED IT!

- **Research**—Identify your vulnerabilities and gather needed contact information.
- **Action Plan**—Outline what needs to be done to mitigate the risks (or respond to emergencies if crisis planning).
- **Communicate**—Educate your employees and train them on the plan.
- **Evaluate**—Assess how you are doing (or for emergency plans, test them out *before* a crisis). Revise as necessary, and do it again.

3.6.1 Crises Can Make or Break Your Reputation

Despite the highly publicized and drawn out legal battles that ultimately led Johnson & Johnson to stop manufacturing baby powder in May 2020 due to concerns over cancer, the company's handling of Tylenol tampering in 1982 will long be lauded as a model for communicating in a crisis. After seven people died from cyanide-laced Tylenol capsules in the Chicago area, the company acted swiftly and decisively, and communicated clearly from the outset their priority in the public's safety—not only in words but actions. Approximately $100 million worth of product was voluntarily pulled from the shelves across the nation and production and advertising screeched to a halt. Post-crisis, Johnson & Johnson set another bar, this time in product quality. The company introduced a new tamper-resistant packaging that soon became the industry standard. They took a substantial hit in the marketplace but the business fully recovered and prospered. Johnson & Johnson was widely praised for its response, including by the media.

On the opposite end of national cases studies in crisis management lies BP. The company will likely be remembered for the five words that slipped from the mouth of its (now-former) CEO *"I'd like my life back"* and the myriad of crisis communication missteps that occurred following the oil rig explosion in the Gulf of Mexico in April 2010.

So, what if yours is not a Fortune 100 company like Johnson & Johnson or BP? Can you manage alright by just winging it? It is possible but not likely. Consider how the odds are stacked against you:

- You may have the wrong spokesperson. (Not everyone makes a good spokesperson, and even fewer do well under pressure without practice.)
- You will not know who to alert and how to alert them.
- You will waste precious time scrambling for resources and facts.
- You will lose your focus while chasing cascading problems.
- Others will have the luxury of shaping your narrative.

Add to this list the potential for technical errors, and the odds of an event careening out of control exponentially increase. The Freedom Industries accidental release in West Virginia offers a good example of how technical unpreparedness can exacerbate communication issues in a crisis. The facility managed a number of chemical storage tanks situated along the Elk River in Charleston. In January 2014, about a week after being acquired through a merger, the facility experienced a leak from a tank containing a coal-washing chemical. The leak contaminated the water supply of approximately 300,000 residents, and sent about 300 anxious people to the emergency room for conditions later determined to be similar to symptoms of colds, flu, and other common viruses.

In short, the technical failures resulted from long-term neglect that led to corrosion of the tank bottom, leaky containment and drainage systems, and a lack of proper maintenance and inspections. The communication failure began with a lack of complete and up-to-date inventory and chemical safety information. Facility representatives were unable to immediately confirm what material was stored in the tank. Not only was incorrect and incomplete information initially communicated verbally to the downstream water authority, it took the facility six days to determine the full composition of the leaked material.

In addition to the chemical characteristics, estimates on the volume of the spill continued to evolve in the ensuing days, causing confusion among responders and authorities, and eroding public trust. Internally, the company had no crisis communication plan in place. When the new owner of the facility arrived on the scene, he was obviously unprepared for the ferocious reporters who swarmed him. The company's reputation suffered so badly that even high-priced PR companies jumped ship shortly after boarding.

Not only did the Freedom Industries spill become memorialized in the annals of crisis communication failures, it also served as a wake-up call to industries and organizations associated with ill-prepared facilities. The company owners, president, and plant manager, as well as the facility environmental consultant, were found guilty of environmental crimes. While the company itself quickly folded into bankruptcy, class-action lawyers found deep pockets in both the chemical supplier and the downstream water company that initially distributed contaminated river water to customers. A $25 million class-action suit was settled with the supplier of the leaked product for failing to adequately warn its customer about storage incompatibilities due to the corrosive nature of the chemical. Another class-action suit for $76 million was settled with the water authority for not adequately preparing for such a leak near its intake.

3.6.2 How Well Are You Anchored for the Storm?

If you accept that it is only a matter of time before a crisis occurs, and if you accept that performing well could vastly improve your image while doing badly can crush your reputation, what will help set you apart from the herd? The answer, according to Fink (2013), is good will and a reputation for being a straight shooter that you have cultivated over the years. Fink writes about these buildups of trust as *reservoirs of good will*. The deeper the reservoir you build, the less chance your ship runs aground.

> "It is the companies that lack a moral compass or that allow themselves to be tossed about on the stormy seas of a crisis, when a firm and sure hand on the tiller is needed, that suffer the most."
>
> *Crisis Communications: The Definitive Guide to Managing the Message, Steven Fink, 2013*

In *The 7 Habits of Highly Effective People* (Covey, 1989), this concept is discussed in terms of *emotional bank accounts*. While Covey discusses emotional bank accounts in terms of personal relationships, the concept works just as well in other relationships. You make deposits through being available, open, honest and considerate, and by keeping commitments. You drain your reserves by acting aloof, inconsiderate, and deceptive, and by paying lip service.

If you have invested in local relationships over time, you earn another benefit that is worth its weight in gold—you know the facts on the ground. This is especially important when considering the distortions in public feedback you may face through social media. As an example, in the aftermath of a gas well explosion at a Chevron well site in the small Pennsylvania community of Bobtown in February 2014, a controversy brewed about a gesture made by the company to local residents. The explosion tragically killed a young worker and the resulting fire took days to contain. As the site was not close to homes, residents were not in danger, but were impacted by ongoing traffic restrictions. Chevron recognized both the graveness of the initial loss and the ongoing traffic problem created for residents. Wanting some face time with residents to hear concerns and field questions, Chevron sent representatives door-to-door to answer questions and offer a coupon for dinner from a pizzeria—the only restaurant within 80 miles of the site (CBS Local, 2014). News of the pizza coupons flashed through social media networks at lightning speed. Sarcastic and scornful diatribes blanketed the Internet by angry tweeters and bloggers. News agencies picked up the story and Chevron was even skewered on the national cable show *The Colbert Report*.

Before long, a worldwide petition was circulated demanding that Chevron apologize to the citizens of Bobtown. By the time the petition's organizer was interviewed in March, the petition had 12,000 signatures (CBS News, 2014). At first blush, the social media blitz signaled a devastating failure in crisis management and communication. However, because Chevron had developed a relationship and ongoing dialogue with the local residents, the company knew that the local residents were understanding and appreciative of the gesture. In fact, not one resident of Bobtown signed the petition. The person who started the petition lived 250 miles away from the site on the other side of Pennsylvania. Those who signed included citizens from California, Alaska, Florida, the Netherlands, Australia, Bulgaria, Costa Rica, Germany, and Italy (CBS News, 2014).

LIMITATIONS ON THE TRUST BANK THEORY

While a buildup of good will, trust, and reputation serve both people and companies well, particularly in the long run, crises communication expert Dezenhall (2014) points out that the trust bank theory has limitations in crises. In some cases, says Dezenhall, the squeakier-clean the organization's reputation, the harder a company may fall in a crisis. Points out Dezenhall, "...there is a deep impulse to step on the feet of the kid wearing shiny new shoes."

3.6.3 Do Not Ignore Social Media

You may view communicating over social media as a boon for business, a necessary frenemy, a luxury you cannot afford, or something you want to avoid like the plague. Even if you fall within the last two categories, however, ignoring social media is not an option. At a minimum, someone in your organization should be monitoring social media to help keep fingers on the pulse. If you are a large organization, someone is likely assigned to do this as part of his or her job. If you are a small organization with no social media presence or resources, there are simple and affordable options for doing this. For instance, Google Alerts is a popular and free application that monitors the web for key names and words, and sends you emails with links to the hits.

Twitter can be an excellent tool for communicating in an emergency. Within seconds, you can reach multiple audiences, without the filter of the media, and in your own words. The social media app is not without its limitations and risks, however. As in all communication realms, social media practice is best guided by core principles and strategic planning.

Another bare minimum in the social media department is a social media policy. Even if you do not use social media as a company, your employees do. If you do not establish and communicate a policy to them regarding when and how they may (or may not) act in your name, you risk an unnecessary crisis. Even well-meaning employees can inadvertently cause embarrassment—whether approved or not to speak on your behalf. And embarrassing tweets are memorialized forever.

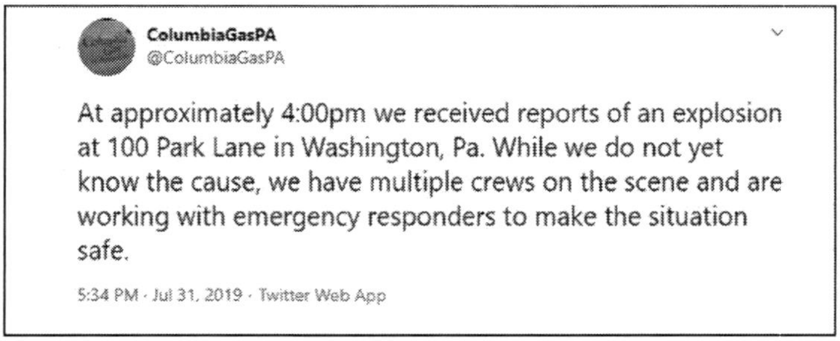

To tweet or not to tweet. Twitter can be a valuable tool for communicating in a crisis.

While Twitter and Facebook can be valuable tools in emergencies, users must be comfortable relinquishing some control in the conversation space. Social media platforms were created to engage and grow networks, and so companies like Twitter and Facebook are slow to infringe on this quality even as they address its inherent problems. Twitter in particular has drifted increasingly toward being a virtual Colosseum since the app was launched in 2016. As of the beginning of 2020, the platform had been tweaked a bit to help limit some of the on-screen bloodshed. For instance, features were added to allow users to hide replies on their feeds and to only allow replies from those who are following the tweeter.

Social media applications will continue to change quickly and often, so strategy and policies based on individual business principles are critical at the outset. It is likely that these types of guiding policies led Johnson & Johnson to engage with online critics about concerns over potential asbestos fiber in talcum powder in 2018, and led Wayfair *not* to respond on Twitter to critics during a politically-motivated employee walkout in 2019. Both approaches are fine as long as they demonstrate consistency with corporate principles and policies.

3.6.4 Write, Test, and Revise

Few things are needed more, yet so consistently overlooked, than a crisis communication plan. Crises communication plans should be well thought out, simple to follow (some experts recommend one laminated page), and tested regularly. All plans rely on some form of vulnerability audit or risk assessment to identify what could go wrong. (The evaluation should explore disruption possibilities beyond accidents and natural disasters, such as employee sabotage.) Additionally, all plans require designation of a crisis communication team and spokesperson, as well as notification lists and pre-approved statements. Table 3.3 provides examples of information that may be included in a crisis communication plan.

TABLE 3.3

Example of Information to Include in Crisis Communication Plan

Example of Information to Include in Crisis Communication Plan
Emergency Communication Policy and Procedures
Crisis Communication Team
Stakeholders
Communication Channels
Communication Monitoring
Scenario Action Plans (from risk assessment/vulnerability audit)
Attachments/Inserts/Stand Alones:
Emergency Communication Forms/Worksheets
Call Tracking Forms/Contact Logs
Talking Point References—Immediate Responses to Media
Master Contacts Lists

Avoid the temptation to make your crisis communication plan so thorough that it becomes unworkable. In some cases, it may make sense to focus on a few more likely scenarios and then maintain information for other scenarios as an appendix or other accessible document. As with other EHS plans, the one element of your crisis communication plan that will change often is your contact list. Make sure you update it at least once per year and share your updates with emergency planning authorities.

A crisis communication plan that accumulates dust is about as helpful as having no plan. Whether table-top drills or mock incidents, be sure to test your written plan once it is prepared and regularly from then on. It is the only way to work out the bugs and create a better fit for your organization. If your spokesperson has not been through media training, it is a good idea to video tape a mock news interview for practice. If your EHS department conducts regular emergency drills, piggybacking crisis communication exercises creates a more realistic practice session for both parties and saves the effort of coordinating two separate training events.

3.7 Invest in Being There Early

Despite the abundance of evidence that being unprepared for sound communication—whether in an emergency or not—can have devastating consequences, many organizations are still slow to fold risk communication into their strategic plans and budgets.

Be first, be right, be credible is not just useful advice for crisis communications. On a grand scale, consider the implications that not being there early had for the chemical industry when activists pushed to ban chlorine, or the energy industry with the current push to ban fracking

During a Crisis:

- Show remorse for injuries and loss of life.
- Have a genuine concern for the incident.
- Stay calm, remember fundamentals.
- Put your crisis plan into effect as soon as possible.
- Communicate with your employees.
- Stress what is being done to mitigate the problem.
- Tell the truth.
- Do not place blame or speculate on a cause.
- *Always remember*—actions speak louder than words.

(more on these issues in Chapter 5). These stories play out in small scale every day for organizations.

3.8 Do Not Go Halfway

A common mistake for organizations that are first warming to the idea of investing in open, honest, and transparent communications is to go halfway. This often occurs after projects fail the first time due to poor stakeholder communication and relationships. Compared to slogging on under burgeoning consultant and attorney fees, playing a little softball might seem like a good option.

The problem with this situation, however, is that the change in course is usually more a tactic than a principle. As soon as the going appears to get tough, organizations pull back out of fear that the row will be just as difficult to hoe, and that unnecessary concessions will be left on the table. In these circumstances, companies would be better having stayed the original course. Going halfway and then turning back sends the signal that you are simply being manipulative. Stakeholders will feel betrayed and less likely to trust you next time around. As a result, you will be left with more expense *and* a buildup of bad will waiting for you the next time you need to come to the table.

If you want to help strengthen your company's resolve in principled communication, identify the potential barriers that exist internally and work to address them. Some of these barriers may include:

- Mindset that communication and outreach are merely fluff and public relations.
- Disconnected functions within the organization.
- Tendency of upper echelon management to tightly control information.

In *Risk Communication: A Handbook for Communicating Environmental, Safety, and Health Risks*, (Lundgren & McMakin 2018) the authors shared the results of a study by Caron Chess at Rutgers University Center for Environmental Communication that showed organizations are more successful at communicating risk if they possess the following characteristics:

- "The organization has a mechanism for upward flow of information.
- The organization has a diffraction of responsibility (communication is everyone's business as opposed to being within the sole purview of a public affairs function).
- The organization has a permeable boundary—there are numerous ways for the community (or audience) to get information about the organization's activities."

3.9 Temper Your Expectations

Communication practitioners cannot guarantee that if you follow sound communication principles and planning that you will emerge unscathed from a communication effort or crisis. They can, however, guarantee that if you ignore the recommended principles and planning, you are much more likely to damage or destroy your organization, whether now or later.

Business climates and circumstances continue to change, and increasingly faster than ever before. If you are guided by a moral compass and a steady hand, you are far more prepared to weather the storm. Moreover, in ways that can be difficult to measure, sound planning and principled practices will continue to build your reservoirs of trust and strengthen your connection with the stakeholders who will, at one point or another, be passing judgment.

CASE IN POINT

Investments Pay Off in Dividends

Approximately one year after initiating a public communication and outreach program, a chemical company experienced an accidental release of an extremely toxic substance. As a precaution, local responders evacuated about 100 residents. Because the company had already been openly communicating about its EHS and emergency preparedness programs, and had been proactively training with local responders, their actions before, during, and after the crisis were principled, effective, and consistent. In addition to following up with local authorities, first responders, neighbors, and other stakeholders, the company also wrote about the incident in the next issue of its quarterly community newsletter. In addition to explaining what happened and what new safeguards were put in place as a result of the incident, the company boldly reinforced its commitment in this way, "We will not manufacture this product again until we know that all safeguards put in place as a result of the investigation are working." What could have been a chaotic and lawsuit fraught outcome was instead a demonstration of competency and principle.

References

CBS Local. 2014. Chevron Controversy: Free Pizza Coupons After Gas Well Explosion. https://pittsburgh.cbslocal.com/2014/02/18/chevron-controversy-free-pizza-coupons-after-gas-well-explosion/ (accessed February 1, 2014).

CBS News. 2014. Pizza coupons "scandal" blown out of proportion, town's locals say. Accessed March 7, 2014. https://www.cbsnews.com/news/chevrons-pizza-coupons-no-scandal-bobtown-pennsylvania-locals-say/

Covey, S. 1989. *The 7 Habits of Highly Effective People.* Rev. Ed. New York, NY. Simon & Schuster Fireside.

Dezenhall, E. 2014. *Glass Jaw: A Manifesto for Defending Fragile Reputations in an Age of Instant Scandal.* Rev. Ed. New York, NY: Twelve.

Fink, S. 2013. *Crisis Communications: The Definitive Guide to Managing the Message.* Rev. Ed. New York, NY: McGraw Hill.

Lundgren, R., and A. McMakin, 2018. *Risk Communication: A Handbook for Communicating Environmental, Safety and Health Risks,* Sixth Edition. Hoboken, NJ, John Wiley & Sons.

Project Management Institute, Inc. 2013. The High Cost of Low Performance: The Essential Role of Communications. https://www.pmi.org/learning/thought-leadership/pulse/-/media/pmi/documents/public/pdf/learning/thought-leadership/pulse/the-essential-role-of-communications.pdf?v=e1f0e914-4b3a-456f-b75e-40101632258b&sc_lang_temp=en

4

Tailor Your Approach and Messages to Your Audience

Introduction

Unless you are communicating solely to coworkers who are performing similar jobs, your audience is not a homogenous group of people who share your perspective. Your audience, and even *the public,* will be made up of groups and individuals with a variety of backgrounds, experiences, interests, and agendas. To be effective, your communication approach and messages should be tailored accordingly.

4.1 Identify Your Audience

As with every element of design in your communication effort, identifying your audience begins with a review of your purpose and objectives. The first question to ask yourself is *Why are you communicating?* For example, maybe you are engaging stakeholders in a public participation process required for a new major environmental operating permit application. In this case, you may be meeting the public participation regulatory requirements of the permit process in a manner that is consistent with, and builds upon, your long-term community outreach program.

Once you establish the *why* you are communicating, you can move on to the *who.* For instance, perhaps the regulatory process requires notifications to residents within a half-mile boundary of the facility—that is one group of stakeholders that will make up your audience. The regulations (as well as common sense) may also require notification of municipalities and county governments within or near the facility. If you have an established outreach program, perhaps you have a community advisory panel whose members should be included on the list, along with any civic or environmental organizations already on your mailing list. If the permit application involves a process related to local right-to-know reporting, first responders should be included, and so on. Local reporters might also be among your potential audience members. As with any communications effort, employees and labor union leaders should always be included as an audience group. (If it makes sense for your effort, you can further group employees into subsets.)

Avoid the temptation to exclude groups that you expect to be adversaries on the issue. If they have to seek you out, they will feel more compelled to weigh in as the opposition. It may feel counterintuitive, but including them up front actually invites a more measured response. Note that this recommendation only applies to those stakeholders who have a history of, or high likelihood of, involvement in your project or program; you do not want to search for new opposition parties who would otherwise not have known about, or held interest in, your effort.

How and when you communicate to these various groups that make up your audience will be flushed out in your plan.

Questions to ask to help identify stakeholders:

- Who needs to be involved?
- Who is likely affected?
- Who is likely to perceive they are affected?
- Who is likely to be upset if not involved?
- Who has previously been involved in this or related issues?
- Who could help ensure you receive a balanced range of opinions on the issue?
- Who might provide third-party support?
- Who would you least like to communicate with on this issue?
- Who is most active, attentive, or passive?

Adapted from A Risk Communication Primer—Tools and Techniques, Navy and Marine Corps Public Health Center, Environmental Program, undated.

4.2 Prioritize Your Audience

As shown in Figure 4.1, public stakeholders fall into one of three categories: Opponents, supporters, and in-betweeners—those who are concerned or could become concerned. Stakeholders who are opposed are often

FIGURE 4.1 Three categories of stakeholders.

motivated by preset agendas and emotions that make them uninterested in constructive dialogue. Thus, the extent of this group's influence on your communications effort will be how other groups perceive your treatment of them. You must still share information with them, listen to their concerns, invite them to meetings, and treat them with courtesy. If you do not, other stakeholders may be more sympathetic with them and lose some trust in you.

Those who agree with you generally give you a large buffer in your trustworthiness and credibility. The extent of this group's influence on your communications effort will be how others perceive their trust in you as well as the insight they may offer with regard to other stakeholders. Thus, you will want to maintain a positive relationship with this group, seek their feedback, and ask their advice.

Stakeholders who are concerned or could become concerned will comprise the greatest portion of your potential audience, and also the greatest potential for productive dialogue. This group includes the whole spectrum of uninterested to very interested individuals, all of whom could move up and down this spectrum. This collective group should be the primary focus of your risk communication efforts, with particular emphasis on those known to be concerned. You will want to ensure you meet the information needs of this group and avoid giving its members any reason to move into the opposition group.

4.3 Analyze Your Audience

When it comes to understanding your audience, more information is always better, and data trumps assumptions. However, the depth of analysis you can afford will depend on the resources and time at your disposal.

At a minimum, non-intrusive analyses can be accomplished fairly quickly to include an examination of news coverage, general community demographics, prior communications and interactions, social media accounts of opinion leaders and activists, and information on active organizations and interest groups. This information will provide a basic understanding of your audience.

Where you can afford the time and resources, and where in-depth information is needed, you can supplement basic audience analysis by conducting interviews, surveys, focus groups, and similar assessments. The breadth and formality of these information tools will vary based on the project. For instance, you may opt to telephone a few opinion leaders for an informal conversation or you may hire a local university to perform a statistically significant telephone poll.

In performing a more in-depth analyses, seek to better understand the perspectives and concerns of your audience members, learn their preferences for how and when to receive and share information, and identify the gaps between their perceptions and the information you wish to convey. Specific questions and discussion topics will vary based on your communication goals and objectives. For instance, returning to the previous example of obtaining an environmental permit, perhaps your communication objectives are:

- Exceed public participation regulatory requirements for the environmental permit application.
- Demonstrate the company's commitment to protect the environment and human health.
- Enhance the company's reputation as an asset and source of pride for the community.

- Build trust among area residents by being proactive and promptly responding to concerns and requests for information.
- Win acceptance from the regulatory agency on the technical approach in the application.

Given these objectives, you would be interested in knowing which sources of news are most used by stakeholders (many permit applications may still require newspaper advertising, so this must be included), what meeting areas are most convenient, what stakeholders know about the company property and operations, how they feel about the company, what type of community assets are most valued, if they have had prior interactions with anyone from the company, etc.

Table 4.1 shows how basic information learned during audience analysis is used to shape communication efforts and messages.

TABLE 4.1

Using Audience Analysis Information to Tailor Risk Messages

Information Learned	How to Tailor the Message
Audience unaware	Use graphic method—high color, compelling visuals, and theme.
Audience apathetic (or feels like victims)	Open risk assessment and management process to stakeholder participation. Show where past interactions have made a difference. Provide choices.
Audience well informed	Build on past information.
Audience hostile	Acknowledge concerns and feelings. Identify common ground. Open risk assessment and management process to stakeholder participation.
Audience highly educated	Use more sophisticated language and structure.
Audience not highly educated	Use less sophisticated language and structure. Make structure highly visible, not subtle.
Who the audience trusts	Use that person to present risk information.
Where the audience feels comfortable.	Hold meetings in that location.
The method by which the audience gets most of its information	Use that method to convey your message.
Who makes up the audience	Ensure that the message reaches each member.
How the audience wants to be involved in risk assessment or management	If at all possible, given time, funding, and organizational constraints, involve the audience in the way they want to be involved.
Misconceptions of risk or process	Acknowledge misconceptions. Provide facts to fill gaps in knowledge and correct false impressions.
Audience concerns	Acknowledge concerns and provide relevant facts.

Source: Lundgren & McMakin, 2018, reprinted with permission.

As an illustration on the importance of audience in crafting communications, the Centers for Disease Control and Prevention's website on COVID-19 featured hundreds of COVID-19 guidance documents and print resources. Not only was "audience" one of the key cataloging features of the resource listing, it was also the only filtering function available for sorting key-word search results.

4.4 Develop Your Key Messages

After you have learned about your audience, you are set to determine the key points you wish to convey in your communications. Key messages help you to:

- Prioritize and define information.
- Ensure consistency, continuity, and accuracy.
- Stay focused when speaking with stakeholders (including the media).
- Minimize confusion in a crisis.
- Allow for the repeatability needed for information to break through.

In determining what your key messages should be, consider the following:

- What do *you* want to tell your stakeholders?
- What do *they* want and need to know but do not?
- What is likely to be misunderstood?
- What would you put in your opening statement at a presentation or press conference?

Risk communication experts and practitioners generally recommend developing three key messages. A particularly useful framework for developing key messages is what risk communication expert Vincent Covello (2002) terms message mapping. As shown in the public health West Nile Virus communication example in Table 4.2, message mapping involves organizing key points into three main messages. Supporting information for each main message is then also layered in groups of three.

Media training expert Brad Phillips suggests making key messages one-sentence statements that incorporate two items: *Your* most important point and the *audience's* most important need or value (Phillips, 2013). He also recommends that supporting information for each key message be grouped into categories of stories, statistics, and sound bites. Providing two statements for each provides a mix of supporting information that can be used in

TABLE 4.2

Sample Message Map

Stakeholder: General Public.
Question or Concern: What can people do to prevent West Nile Virus?

Key message 1	Key message 2	Key message 3
Remove standing water.	Wear protective clothing.	Use insect repellent.
Supporting fact 1-1	**Supporting fact 2-1**	**Supporting fact 3-1**
Remove old tires which collect water and serve as breeding grounds for mosquitoes.	Wear long sleeved shirts.	Repellents containing DEET are recommended.
Supporting fact 1-2	**Supporting fact 2-2**	**Supporting fact 3-2**
Empty or clean flower pots and bird baths daily.	Wear long pants.	Use 23% DEET.
Supporting fact 1-3	**Supporting fact 2-3**	**Supporting fact 3-3**
Empty and clean cat/dog water bowls daily.	Especially at dawn and dusk.	Do not use repellents that do not contain DEET.

Source: United States Environmental Protection Agency, EPA/625/R-06/012, August 2007. https://nepis.epa.gov/Exe/ZyPURL.cgi?Dockey=60000IOS.txt.

interviews and other conversations to support key messages without sounding too repetitive.

If you are developing key messages for individual projects, they should be consistent with, and reinforce, your overall company mission and messages. Effective key messages are (Weatherhead, 2011 and Covello, 2002):

- *Concise*—If possible, limit each key message to 9 words, or 27 words for all three. If possible, each should be capable of being spoken in 3 seconds, or 9 seconds for all three.
- *Simple*—Use easy-to-understand language; avoid jargon and acronyms. In general, keep messages at a middle-school level of readability for audiences that include the general public.
- *Strategic*—Define, differentiate, and address the benefits of your position.
- *Relevant*—Balance what you need to communicate with what your audience needs to know.
- *Compelling*—Design meaningful information to stimulate action (particularly for precaution/care advocacy).
- *Memorable*—Ensure that messages are easy to recall and repeat; avoid long, run-on sentences.
- *Real*—Use active voice, not passive; do not use advertising slogans.
- *Tailored*—Adjust language and depth of message for different target audiences.

Challenges surrounding facemask recommendations in coronavirus risk communication. Conflicting advice among scientists and health experts can both erode people's trust and reduce the effectiveness of safety advice. In the case of coronavirus (COVID-19), initial federal government recommendations against wearing surgical masks in public appeared to conflict with some recommendations made by the World Health Organization, practices in other countries, and the results of previous flu protection studies published by the CDC. Coupled with the government's claim that healthy citizens do not need masks in public was the request to not purchase masks because of the severe shortage for healthcare workers, another seeming contradiction. A university professor's scathing editorial in the New York Times (Tufekci, 2020) memorialized the confusion and negative reactions likely experienced by many members of the public over the facemask advice. Months after the initial advice was given, federal authorities reversed their recommendations on wearing masks in public.

Once you prepare a draft of the key messages, test those messages to ensure that they resonate with internal and external audiences. If possible, this testing should include people in your target audience. Testing messages allows you to:

- Increase your chance of success in conveying information.
- Assess the comprehension and recall of your messages.
- Identify strengths, weaknesses, and potentially sensitive or controversial language.
- Identify cultural impacts.
- Assess personal relevance.

Based on the results of the message testing, revise and finalize the key messages. Over time, routinely revisit the key messages to ensure that they still meet your needs and those of the audience. Key messages should also continue to reflect current trends, research, and issues your organization is addressing.

Once your messages are established, they should serve as the backbone of your communication effort. Every piece of communication collateral and every discussion should be built around them.

4.5 Stick to Your Messages

As stated previously, key messages serve as the backbone of your communication effort. They should keep you connected to your project objectives and to your audiences' needs. Thus, every plan you create, action you take, and statement you make should be aligned with your messages. If any of these is out of line, you should determine if your objectives or messages are off target.

Sticking to your messages for your communications project means drawing on them for your discussions and information materials. While they will help keep you grounded, they are not meant to be talking points that are simply repeated verbatim in place of dialogue.

For nearly every communication effort undertaken, you can anticipate most of the questions you will receive. Drafting responses to these anticipated questions to share among your team is helpful in maintaining

BRIDGING STATEMENT EXAMPLES – GET BACK TO YOUR MESSAGES WHEN SPEAKING WITH THE MEDIA

Spokespersons should use statements such as the following to return to key messages or to redirect when discussions with journalists move off course. For a more complete list of bridging statements, see Appendix D.

- "What matters most in this situation is ..."
- "Let me put all this in perspective by saying ..."
- "Before we continue, let me take a step back and repeat that ..."
- "This is an important point because ..."
- "Let me just add to this that ..."
- "I think it would be more correct to say ..."
- "In this context, it is essential that I note ..."
- "Before we leave the subject, let me add that ..."
- "While...*X*...is important, it is also important to remember ...*Y*..."
- "It's true that...*X*...but it is also true that ...*Y*..."

Adapted from "Effective Risk and Crisis Communication during Water Security Emergencies: Summary Report of EPA-Sponsored Message Mapping Workshops" EPA/600/R-07/027March 2007.

alignment with messages and consistency in communications. In cases where responses must be approved by attorneys, ensure that the answers are preapproved. (Press your legal team to avoid unnecessarily constrained responses.) Finally, do not leave off the list any uncomfortable questions that you hope are never asked—those are the most important of all for preparation. More information on question-and-answer documents is provided in Section 9.2.2.

Sticking to your messages for media interviews requires a particular skill called bridging. As discussed in Chapter 8, effectively engaging with the news media means recognizing that their agendas do not align with yours. Journalists view themselves as independent fact-finders and storytellers who are obligated to present opposing views. Your company spokesperson must represent your interests. In doing so, he or she must be prepared to steer back to key messages while answering questions.

CASE IN POINT

If You Do Not Understand Your Audience, You Will Lose Them

A company president speaking at a public meeting about his facility's controversial environmental permit application did not understand the interests of concerned residents gathered in the audience. He addressed the residents as if they were potential clients. Instead of discussing aspects of the project that worried them, he spoke about the company's technology, new service areas, and plans for development. "Who cares?" was the mildest of the various comments that could be heard in the back of the room as residents talked among themselves during his presentation. The president's total lack of consideration for the residents' perspective deepened their resolve to oppose the project.

References

Covello, V. 2002. Message Mapping, Risk and Crisis Communication. Paper presented at the World Health Organization Conference on Bio-terrorism and Risk Communication, Geneva, Switzerland.

Lundgren, R., and A. McMakin. 2018. *Risk Communication: A Handbook for Communicating Environmental, Safety and Health Risks*, Sixth Ed. Hoboken, NJ: John Wiley & Sons.

Tufekci, Z. 2020. Why Telling People They Don't Need Masks Backfired. New York Times Opinion. March 17. https://www.nytimes.com/2020/03/17/opinion/coronavirus-face-masks.html.

Weatherhead, D. 2011. Key Message Development: Building a Foundation for Effective Communications. http://prsay.prsa.org/2011/12/02/key-message-development-building-a-foundation-for-effective-communications (accessed April 23, 2020).

5

Address Emotions before Facts

Introduction

On its surface, communicating about environmental and health risks may feel challenging because of its reliance on sophisticated science. Certainly, sophisticated science adds a degree of difficulty to communicating. However, the greatest pitfalls in communicating about environmental and health risks are related to simple human nature. Stakeholders' reactions to risk communication, like any form of communication, are largely driven by gut instincts and feelings. What follows is advice to help you navigate the emotional forces that can make or break your effort. If you do not address these first and foremost, the rest may not matter much.

5.1 Accept That Facts can be Irrelevant

We accept the truism in politics, family discourse, and many other aspects of our lives—facts may have little to do with people's positions. Yet, when it comes to soliciting input on matters involving environmental and health risk, we expect others to check human nature at the door. And why not? We do not count emotions in calculating safe loads for bridges, formulating chemical compounds, or designing protective equipment.

> "All the studies show that, even with good risk communication, people carry on doing what they did before."
>
> *Professor David Spiegelhalter, Quartz, Rathi, 2016*

The crux of the matter is that most people generally trust the technical credentials of those doing the calculating and planning. What is left are all the unspoken value components that come into play—control, fear, fairness, respect, and so on. On top of that, stakeholders often approach issues with very different experience and knowledge bases. Thus, facts can become irrelevant because stakeholders are too upset to care about them or because their understanding may not be sufficiently informed to put facts into context. And for stakeholders with personal agendas, facts simply do not make it into the equation.

Accepting that facts can be irrelevant can help temper your expectations and prepare you for meeting the challenges most likely to derail your communication. A number of risk communication experts have addressed the non-technical forces that influence stakeholders' responses to environmental and health risks. Two prominent views that give context to the issue are Peter Sandman's Risk = Hazard + Outrage formula (Sandman, 2010) and the "Mental Models" approach put forth by Carnegie Mellon University researchers (Morgan et al., 2002).

5.1.1 The Outrage Variable

If you have done any Internet surfing on risk communication, you have likely come across the name of Peter Sandman and his trademark conceptual formula "Risk = Hazard + Outrage." What Sandman emphasizes with this formula is that experts see risk very differently from the general public. In fact, when measured statistically, the correlation coefficient for risk rankings between lay members of the public and scientists is extremely low – 0.2 (1.0 represents a perfect correlation) (Sandman, 2006).

Experts define risk as some combination of the hazard itself and the likelihood of it happening. There is no emotion in the formula. However, people who perceive that they could be affected by a hazard will weigh things very differently. Their formula is heavily weighted by factors that have nothing to do with science or risk or chance. Rather it is a combination of all the psychological factors that may make them upset or apathetic or somewhere in between. As a result, there are many risks that make people furious even though they cause little harm, and others that kill many, but do not make anyone angry. These non-technical "outrage" factors have been identified by many risk communication experts with little variation over time. A compilation of them is presented in Table 5.1.

Generally, public concern can be plotted on a hazard verses outrage quadrant (see Figure 5.1). As shown in the quadrant, risks that are actually low in hazard can still be associated with high outrage. The same is true for hazards that are actually high—they do not necessarily correlate with high outrage. If both outrage and hazard are high, you will probably find yourself in the quadrant where you are dealing with a crisis. If outrage is low, but the hazard is high, you are probably dealing with a situation where you are trying to motivate people to take risk more seriously. If you have low hazard but high outrage, you are probably dealing with a situation where you are trying to calm people down.

TABLE 5.1

Factors Affecting Audience Reaction to Risk

Seen as Less Risky	Seen as More Risky
• Voluntary	• Involuntary
• Individual control	• No or little control
• Fair	• Unfair
• Naturally occurring	• Created by humans
• High trust in source	• Low trust in source
• Familiar	• Unfamiliar
• Affects everybody	• Affects children
• High personal benefit	• Low personal benefit
• Chronic	• Catastrophic
• Positive experience	• Negative experience
• Morally irrelevant	• Morally relevant
• Not memorable	• Memorable
• Responsive process	• Unresponsive process
• Transparency	• Secrecy

Examples of how non-technical factors impact people's reactions to risk abound in daily life. Imagine, for instance, the difference you feel when you are driving a car versus sitting in the passenger's seat with an equally skilled driver at the wheel. In this case, the amount of individual control you have as the driver greatly impacts your comfort level.

Reactions may even differ when the same hazard is presented in a different setting. For instance, Pennsylvania is home to a geology rich in elements that degrade into radon progeny. Homes within the Commonwealth have some of the highest radon levels in the United States. While natural background

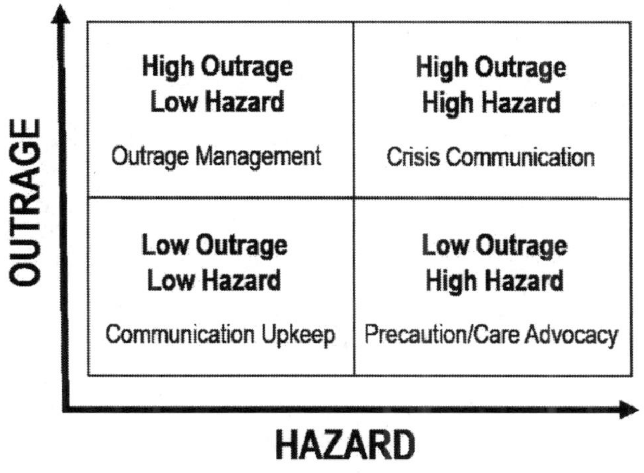

FIGURE 5.1 Relation of outrage and hazard quadrant to categories of risk communication.

levels of radon in Pennsylvania range from 0.2 to 0.7 picocuries per liter (pCi/L) (King & Spalding, 2015), concentrations as high as 6,176 pCi/L have been found inside homes in the Commonwealth (Upper Saucon Township, 2016). Radon is a leading cause of lung cancer, second only to smoking. Yet, despite numerous awareness campaigns and public service announcements that blanket the airwaves every winter, many people still do not test their homes. Comparatively, residents near municipal and residual waste landfills that accept oil and gas "fracking" waste have attended meetings to express outrage at the potential for radon exposure. This is despite the fact that studies have documented that radon concentrations at and around these landfills lie within naturally occurring background levels (Pennsylvania Department of Environmental Protection, 2016). The "outrage" factor causes perceived risks to far outweigh real risks, and vice versa. The factors affecting reaction to risk are clearly different. Radon in the home is seen as a voluntary risk, under individual control, in a familiar environment, with no apparent or acute health effects, and with a personal price tag for testing and remediation. The opposite is true for the landfill, which is seen as an involuntary exposure, out of residents' control, an unfair and unfamiliar problem created by corporations with no benefit to residents, and no cost to residents for stopping the perceived "exposure."

Understanding the factors affecting audience reaction to risk allows you to plan for potential outrage and determine which factors you can influence to avoid unnecessary and unhelpful reactions.

5.1.2 The Impact of Mental Models

Researchers at Carnegie-Mellon University (Morgan et al., 2002) advocate approaching risk communication with the use of "mental models." This approach stresses the importance of analyzing what people know and how they think about an issue before trying to communicate about it.

In *Risk Communication, A Mental Models Approach*, (Morgan et al., 2002), the authors aptly point out that people communicating about environmental and health risks often miss the mark because they do not focus on the things that people need to know but do not already. Rather than learn what people believe and what information they need to make decisions, communicators typically ask technical experts what they think people should be told. Thus, they are likely to miss the mark—resulting in confusion, alarm, or disinterest on the part of the audience.

The mental model approach involves a five-step process to create and test risk messages:

1. *Create an expert model*—This entails reviewing current scientific knowledge about the processes that determine the nature and magnitude of the risk, and then summarizing it from the perspective of what can be done about it. This results in the creation of an influence

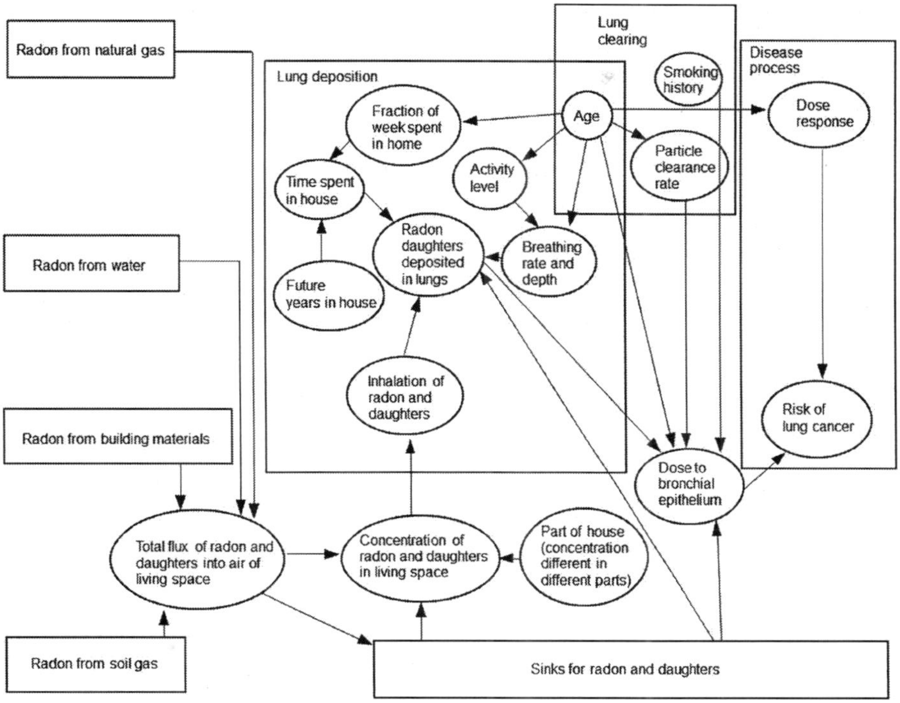

FIGURE 5.2 Abbreviated expert mental model for health effects of radon. (Reprinted [adapted] with permission from W.G. Morgan, B. Fischhoff, A. Bostrom, L. Lave, et al., Communicating Risk to the Public: First, know what people know and believe. *Environmental Science & Technology, ES&T Features, Vol. 26, No. 11, 1992, 2051*, American Chemical Society.)

diagram that is reviewed by experts other than those who created it. The expert model is an attempt to pool together everything known, or believed, by the community of experts that is relevant for the risk decisions people face. Figure 5.2 shows an abbreviated expert model created for the risk of radon in the home.

2. *Conduct mental model interviews*—These are open-ended interviews of representatives of the intended audience to learn people's beliefs about the hazard, expressed in their own terms. The interviews should be designed to touch on all topics covered in the diagram. These responses are analyzed in terms of how well these mental models correspond to the expert model captured in the influence diagram.
3. *Conduct structured initial interviews*—These are confirmatory questionnaires that capture beliefs expressed in the open-ended interviews and the expert model. It is administered to larger groups that are representative of the intended audience. What is key here is that

they are open ended, and the interviewer does everything possible to not give information or hints to the person being interviewed. For instance, with regard to radon, an interviewer may say "Tell me about radon." Then, for each concept that respondents mention, the interviewer can later ask them to elaborate. For example, "You mentioned that radon can come into the house through the cellar. Tell me more about that." That way you can develop their mental models without influencing them.

4. *Draft risk communication messages and materials*—Based on the results of the interviews and questionnaires, along with analysis of decisions people face, you then determine where the knowledge gaps and/or misunderstandings are and draft your communication to address that.
5. *Evaluate risk communication messages and materials*—Before finalizing the risk communication, you test it on individuals from the intended audience, using one-on-one read-aloud interviews, focus groups, closed-form questionnaires, or problem solving tasks. This process can be repeated until the communication is understood as intended.

In the radon example used by Morgan et al., evaluation of the mental models and interviews revealed various misconceptions deemed important for risk communication materials. For example, many people believed that radon can permanently contaminate a house. If this belief is not addressed in risk communication materials, audiences may be less inclined to address the issue because they feel there is nothing that can be done anyway.

Obviously, the mental model approach can require more time and resources than most have at their disposal. But it should be a goal to take these steps to the extent possible. For instance, do not just ask experts about what you think people need to know. Look for ways to learn about what your audience actually knows. When you are drafting a communication piece, do not run it by your colleagues who share the same point of view. At a minimum, even under a rush situation, you could get some feedback from someone in another department or completely different job function in your department. In the case of understanding your audience's perceptions before drafting messages and communication material, something is always better than nothing.

5.2 Understand That You Will Be Treated Differently

As an industry representative, you are often seen as the Goliath to the David of ordinary citizens and even of aggressive environmental advocates.

In *Staking Out the Middle* (Sandman, 2010), risk communication expert Peter Sandman explains that the public forgives the distortions and exaggerations of activists because they are seen as serving the public good. They are calibrated to be oversensitive, just like smoke alarms. Industry, on the other hand, is seen as being calibrated too far in the other direction; thus, reassurances are met with skepticism. To complete the smoke alarm analogy, false alarms are not very upsetting, but the failure to sound during a real emergency can be fatal.

Put another way by Berman and Company during a June 25, 2014 presentation at the Western Energy Alliance Annual Meeting, activist groups "have no natural enemies" (Song, 2014). That cannot be said for many industries.

Communicating Your Value to Society is Important. The World Chlorine Council, a global network of trade associations and companies representing chlorine as an international outgrowth of the American Chemistry Council, consistently communicates about the societal benefits of chlorine products and engages in sustainability programs with the United Nations and other partners. Shown here is a snapshot of the cover for the organization's 2017 progress report. (World Chlorine Council, reprinted with permission through American Chemistry Council.)

Accepting that the public has different expectations for industry than for citizen groups and professional activists will help you avoid wasting valuable time and energy on the element of unfairness. Activists get to play on fear and anger, and exaggerate while doing it. You cannot. In fact, you even will be judged harshly by the truths you leave out and the ones you gloss over. Thus, your communications, besides being respectful even under fire, must openly acknowledge whatever kernels of truth exist on the other side, even when they work against you. Otherwise, you will find it difficult to gain credibility.

That said, industry has a right, an obligation even, to counter mistruths. Otherwise, consistent lopsided stories take root into the public psyche. Consider the public campaign that many environmental activists raged against chlorine and chlorine products beginning in the late 1980s and the early 1990s. In an opinion piece by Independent Commodity Intelligence Services, (Davis, 2009), the authors state "Taking an apocalyptic view of a naturally occurring element seemed to many to be an absurdity. But there was a dawning realization through the early 1990s that unless there was more communication, at many levels, industrial chlorine chemicals could be legislated out of existence." At all levels—local, regional, federal, and some cases, international—industry should be consistently sharing truthful information about its benefits to society and its host communities.

Recently, energy sector industries and the businesses that rely on them have engaged in a differently styled information campaign to counter the activist offensives on conventional energy sources. The campaign, managed under the advocacy organization known as the Environmental Policy Alliance and launched under the project title *Big Green Radicals,* featured a number of edgy ads on billboards along major highways.

5.3 Practice Sound Dialogue Skills

Regardless of whether you are communicating in crisis mode, advocating occupational precautions, or preparing for public participation on controversial permits, when you plan to address the emotions that can drive dialogue, you increase your chance of success.

In Crucial Conversations—Tools for Talking When Stakes are High (Patterson et al., 2012), the authors state "…the root cause of many—if not most—human problems lies in how people behave when others disagree with them about high-stakes, emotional issues." Those who find a way to master these high-stakes, crucial moments can dramatically improve the chances for positive outcomes.

While the authors do not explicitly address environmental and health risk communication, the book's theme directly applies to all individuals and companies genuinely seeking productive dialogues where all parties benefit.

This theme runs through another timeless bestselling book, *How to Win Friends and Influence People* (Carnegie, 1981). In the introduction to this book, Carnegie notes that research by the Carnegie Foundation for the Advancement of Teaching and later confirmed by the Carnegie Institute of Technology, revealed that even in such technical lines as engineering, about 15% of one's financial success is due to one's technical knowledge, and about 85% is due to one's personality and the ability to lead people. He then goes on to describe the various people skills that can dramatically improve the outcome of dialogues and interactions with others at all levels.

In matters of environmental and health risk, deference to the following five emotional motivations can help improve your dialogue: pride, adversity to change, humility (yours), empathy, credibility, and trust.

5.3.1 Pride

No one likes to be told they are wrong, and no one likes to take orders. Similarly, no one likes the feeling that others think they are smarter or more important.

If you tell people they are wrong, will that make them want to agree with you, or strike back? What if you prove, and they accept, beyond a reasonable doubt that they are wrong, will that make them want to admit their mistake and move on or will it cause them to lash out in other ways?

In *How to Win Friends & Influence People*, Carnegie offers quotes from philosophers as far back as Socrates and Galileo on the foolishness of proving others wrong to make your point. In *Getting Past No* (Ury, 1991), the author puts it another way—"Don't confuse getting even with getting what you want."

A number of research studies have provided biological evidence that explains our reactions to information with which we disagree. A study published in the *Journal of Scientific Reports* (Kaplan et al., 2016) explored the neural systems that govern resistance to changing beliefs. In the study, brain scans showed that participants with greater belief resistance had increased activity in the dorsomedial prefrontal cortex (associated with selective attention) and reduced activity in the orbitofrontal prefrontal cortex (associated with higher order cognition like decision making). The study also found that participants who changed their mind more showed less bold signals in the insula and amygdala, areas associated with emotion and behavior.

Patterson et al. (2002) discuss the physical responses that result from this emotional stimulation—the activation of the fight or flight response that occurs when your adrenal glands kick into overdrive. "Countless generations of genetic shaping drive humans to handle crucial conversations with flying fists and fleet feet, not intelligent persuasion and gentle attentiveness."

Fortunately, you have the power to change the course of the interaction. The following tips will help you avoid triggering push-back from others' perceived affronts to pride and ego:

- *Resist the urge to fight fire with fire*—While it is our gut instinct, pushing back only results in stronger force against us. Commit at the outset to be the one who shifts the energy in a positive direction.
- *Genuinely listen to people's concerns*—When people feel that their concerns are not being heard, they dig their heels in further. They repeat their positions, and they get angrier. Always practice empathetic, or active, listening.
- *Focus on the problems, not the positions*—By avoiding the temptation of taking an opposite position, you create an opportunity to explore the concerns that lie underneath. You may ask questions like "Can you help me understand why you want that?"
- *Allow people to save face*—Showing people that they are wrong in a blatant manner may bring temporary satisfaction, but it will cost good will. "'Face' is much more than ego. It is shorthand for people's self-worth, their dignity, their sense of honor, their wish to act consistently with their principles and past sentiments—plus, of course, their desire to look good to others" (Ury, 1991).
- *Give people reasons to come to your side willingly*—Once people understand they are not engaged in a battle of wills, they are more apt to consider mutual problem solving. Sharing of information and concerns under these conditions provides the opportunity to bridge the gap between their interests and yours.
- *Use flexible language*—It is easy for us to inadvertently express "informed opinions" as fact, which can damage credibility and also

> leave us painted in a corner. It is also easy to trigger push back from individuals with more sensitive egos, making it more difficult for them to back down gracefully. Patterson et al. express the importance of talking tentatively. Talking tentatively means softening your language to signal that there is room for discussion and fact finding.

If you have been to contentious public meetings, you may have witnessed the moment that one or more of these practices helped turn the corner. If so, you would have heard phrases such as "Look, I'm not here to complain, but..." or "I'm not looking to shut you down, I just..."

USE FLEXIBLE LANGUAGE

Compare the Responses

Below are two potential responses to a neighbor complaint about odor. The factual information is the same, but the approach and tone differ. Which do you think is more conducive to productive dialogue and relationship building?

A fence-line neighbor emails you about the bad odor she noticed yesterday morning. She exclaimed frustration because this was the second time in as many weeks.

Response Option 1: "We checked our meteorological data for yesterday and the wind was predominantly blowing in a direction opposite your house. Additionally, there were no changes in our normal operation. There's no reason to believe it was from our facility."

Response Option 2: "Thank you for alerting us. We checked our meteorological data for yesterday, and the wind appeared to be blowing in a direction opposite your house for much of the morning. Given this, and the fact that our logs indicated no change in operations, we will need to dig in a little deeper to figure out if we were contributing to the odor you experienced. Are you able to provide any more details, such as exact times and the type of smell?

5.3.2 Humility (Yours)

If you were truly wrong about something, apologize. People will not be ready to move on until you do. The push-and-pull force in dialogue is hard at work in situations where both parties know that one side was wrong. The more you hesitate to admit clear fault or generate excuses, the more stakeholders will fixate on your error. The sooner you focus on what you did wrong, the sooner stakeholders are likely to focus on the reasons why it is not entirely your fault. This cycle even occurs with pampered politicians and celebrities, admittedly, though, on a superficial scale. They get caught in wrongdoing, they issue a public apology, and check into a rehab center. Until they apologize and do penance, they will be hounded by the press and a disappointed public. (Again, there may be exceptions here, particularly where legal ramifications exist.)

Even achievements should be discussed with humility. This is especially true if you resisted change until regulators or activists forced your hand. When industries make an environmental improvement after being forced to do so kicking and screaming, and then try to take credit for being green or proactive, it damages their credibility.

Acting with humility does *not* mean that you should admit fault if you were not wrong. Doing so is equally as disingenuous as not admitting fault. It also weakens your credibility and jeopardizes your negotiating position for future interactions. (Humility is further discussed in Chapter 2.)

5.3.3 Adversity to Change

The human tendency to resist change has been explained by both physics—an object at rest stays at rest—and our primitive brains—which are wired to bring comfort with reinforced habits. This resistance is one of the dynamics behind cognitive dissonance, our tendency to simply explain away our bad behaviors, choices, and circumstances. For example, a smoker might say "I know it's not good for me, but I will gain too much weight if I quit, and that would be worse on my body."

We humans are also wired with other coping methods that help keep us in balance. Confirmation bias helps us tune out or explain away opposing points of view. Optimism bias allows us to believe we are all smarter than the average bear. (Statistically, we *all* cannot be above average, can we?) We also have a limited capacity for fear—at some point, our brain just cannot keep piling on all of the things we have to worry about.

In *Loserthink: How Untrained Brains Are Ruining America* (Adams, 2019), *Dilbert* cartoonist Scott Adams warns that "people are not fundamentally rational when it comes to many of life's biggest questions. Instead…[people make] one irrational decision after another" and then cover their tracks "by concocting 'reasons' after the fact. In other words, we are not so much a rational species as a species that experiences the *illusion of being rational.*"

These emotional responses sometimes lie on the opposite end of the spectrum than we normally associate with risk communication. They can create apathy, which can be as challenging to address as outrage. As an example, it can be very difficult to get employees to wear personal protective equipment that was not required in the past. Here again, spewing facts and threats will not do.

Fortunately, there are things you can do to help break through the apathy barrier.

- *Identify people's predispositions and appeal to them*—Building on people's preexisting opinions, attitudes, values, and expectations, even if unrelated to the issue, can help encourage change. As an example, Sandman (2016) points out that construction workers and others in hazardous occupations resisted wearing hardhats when the regulations initially took effect many years ago. It is believed that part of the resistance was linked to the workers' pride in their courage and competence. Workers felt that hardhat requirements called their courage and competence into question, and needlessly so, since many had never experienced a head injury. According to Sandman, strict rules actually forbidding the use of hardhats in low-risk parts of the worksite turned around the perception; such rules implied that only workers skilled enough and brave enough to work in dangerous places wear hardhats. The change was so successful that these workers came to be called, and called themselves, hardhats.
- *Do not lecture*—Lecturing and scolding people often produces the opposite of the desired effect. In *How to Win Friends & Influence People* (Carnegie, 1981), Carnegie recants a tale of frustration by a safety coordinator from an Oklahoma engineering company. Whenever he came across workers who were not wearing hard hats, he would tell them with a lot of authority of the regulation and that they must comply. As a result he would get sullen acceptance, and often after he left, the workers would remove the hats. He decided to try a different approach. The next time he found some of the workers not wearing their hard hat, he asked if the hats were uncomfortable or did not fit properly. Then he reminded the men in a pleasant tone of voice that the hat was designed to protect them from injury and suggested that it always be worn on the job. The result was increased compliance with the regulation with no resentment or emotional upset. Carnegie also recounts the tale of how Charles Schwab handled a group of steel mill employees ignoring no-smoking rules. Reportedly the group of employees were smoking inside the mill, right under a sign that read "No Smoking." Instead of pointing to the sign and asking if they cannot read, Schwab walked over and handed each one a cigar and said "I'll appreciate it boys, if you smoke those outside."
- *Walk the talk*—Do workers have reason to believe that you are just paying lip service? If they do not see you consistently enforcing or following your own rules, they will not bother either. They probably figure it is only something you want them to do during the safety assessment or when an authority figure is present.

5.3.4 Empathy

As much as we all thirst for others to *get us* or *feel our pain*, empathy is a scarce commodity in our interactions. In a speech delivered in 2017, former President George W. Bush offered these words: "Too often, we judge other groups by their worst examples—while judging ourselves by our best intentions." This proves to be a universal human weakness—unintended fallout from the focus on "me, myself and I" that comes into play in all aspects of our lives. It is a self-inflicted flaw in our ability to see the world through the eyes of others.

Being aware of this thinking pattern can help prepare you for being a more empathetic communicator. When you are being empathetic, you are focusing on the needs of those who have the ability to derail your project or operation. If you are only focused on what you want, on the information you feel you need to get across, you are just meeting your own needs. Yet, whose needs must you meet to move forward?

WHEN CANCER AND OTHER HEALTH ISSUES COME UP IN PUBLIC MEETINGS

At the conclusion of the final public meeting on the decommissioning of a former nuclear fuels facility, a woman came up privately and said she appreciated the work done to investigate the potential impact of the plant during its many years of operation. She went on to say that her daughter had worked for about three years in an office directly across from the facility and later died from cancer. This mother believed the plant operations contributed to her daughter's death.

Oftentimes, concerns like these are not brought up discreetly, but rather are angrily shouted during meeting interruptions. Regardless of how these issues arise, you must bear in mind that these are very real and personal experiences. Your reaction must be delivered with respect and empathy.

It is important to note that you can show genuine empathy while at the same time having different interests. Empathy is neither sympathy nor agreement, and it cannot be faked.

For instance, you can show you care by saying "I'm very sorry to hear about…" You can show compassion by saying "I can't imagine how difficult that must be…" And you can show concern by saying "I assume you asked that question because you care about…, which I also care about." Notice that none of these statements project a feeling that others are right or wrong for feeling how they do.

If you want to improve your ability to empathize, make it a priority to strengthen your active listening skills, as discussed in Section 5.4.

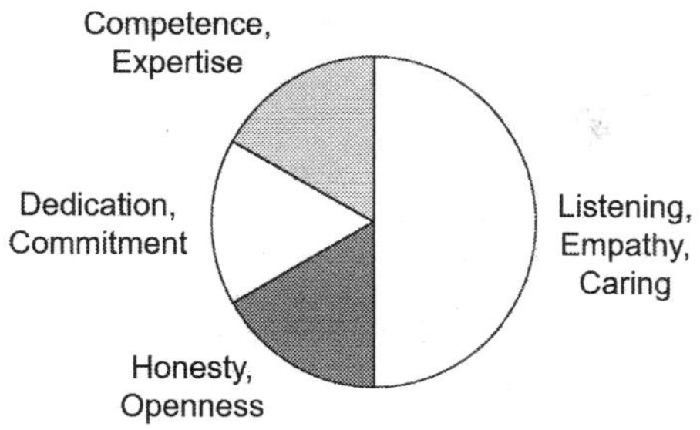

FIGURE 5.3 Stakeholders value empathy most. (Adapted from U. S. Environmental Protection Agency and Covello, 2011.)

5.3.5 Credibility and Trust

Why is trust important? Because it will influence stakeholders' perceptions and their entire interaction with you. If they trust you, they are more willing to listen; and just as importantly, they are more likely to forgive inadvertent mistakes. If they do not trust you, they will be skeptical about everything you say, and will look for reasons to discount your point of view. Consider, for example, how you would treat a small infraction of a friend versus someone you mistrust or do not know.

Covello (2011) often shares an important statistic related to trust factors—50% of the trust others will place in you depends on whether they think you care about them, in other words if you can empathize with them. As shown in Figure 5.3, empathy is more important than your technical knowledge or capabilities, your dedication to your work and even whether you are honest. And just as importantly, if you are new to them, people are likely to decide whether you care within about 30 seconds of meeting you.

Understanding that your ability to show empathy is the most important attribute of trust helps you keep things in perspective when planning to engage in dialogue with stakeholders. No longer should your main, or only, focus be on your technical presentation.

5.4 Commit to Active Listening

Truly listening to what another person has to say is one of the most sincere forms of respect a person can offer. If you attentively listen to people, you are more likely to understand them and thus achieve the empathy that is so important in establishing trust.

The various stages of listening, from not paying attention at all, to being fully engaged, have been described in a number of ways. One the most comprehensive descriptions of these stages comes from Peter R. Garbers *50 Communications Activities, Icrebreakers, and Exercises* (Garber, 2008).

1. *Not listening*—Not paying attention to or ignoring the other person's communications.
2. *Pretend listening*—Acting like or giving the impression that you are paying attention to another person's communications, but in actuality not really paying attention to that individual.
3. *Partially listening*—Only focusing on part of the other person's communication or only giving it your divided attention.
4. *Focused listening*—Giving the other person's communication your undivided attention.
5. *Interpretive listening*—Going beyond just paying attention to really trying to understand what the other person is communicating.
6. *Interactive listening*—Being involved in the communications by asking clarifying questions or acknowledging understanding of the communications.
7. *Engaged listening*—Being fully engaged in communications by listening to the other person's views, feelings, interpretations, values, etc. In engaged listening, both parties are given the opportunity to share and fully express their views, feelings, and ideas.

While we like to think we listen attentively, much of our listening time is actually spent either pretending to be listening or partially listening. If you are scrolling through your phone while someone is talking, you are not listening. If you are not looking the other person in the eyes, you are not listening. If you are thinking about what you want to say while you wait for the other person to stop talking, you are not listening.

If we use active listening to, as Stephen Covey (1989) says, *seek first to understand, then to be understood*, we actually take communication pressure off ourselves. We can kick back, set aside our arguments, and try to grasp what is being put in front of us. Once we do that, people will be more open to listening to what we have to say, rather than taking turns lobbing and ducking arguments.

Active listening takes real practice because it is rarely done. It also takes patience and humility, especially when others are providing negative comments about your operations or being contentious.

If during conversations or presentations you receive valid feedback on your operations that is hard to take, do not get defensive. Instead, "listen, *truly listen*, and try to understand the perspective being offered. Then take ownership of [any legitimate] shortfalls and try to make improvements in the areas of critique you have received" (Willink, 2020). Outsiders' perspectives can reveal errors in your thinking that lead to errors in operations, which, when corrected, can lead to benefits to the company and community.

Sandman (2010) offers excellent examples of providing feedback during active listening for a contentious public meeting:

> "Let me see if I've heard you right...."
>
> "I think some people probably feel that...."
>
> "I wonder if some of what I've heard tonight means that...."

You will note that this language is flexible, or tentative. By echoing what you think you have heard in a tentative manner, you allow room for dialogue where both parties can correct any misimpressions. It also makes it easy for stakeholders to revise or reinterpret their earlier comments without losing face.

ACTIVE LISTENING—DO'S AND DON'TS

DO'S

Give the Speaker Your Undivided Attention

Face the speaker in a relaxed manner and use eye contact. Avoid all distractions, *especially your phone*. If you have time limits, state them in advance instead of checking your watch.

Listen With the Intent to Understand

Avoid coming to conclusions or thinking about how you will respond. Do not interrupt. Ask questions only to ensure understanding, not to support your analysis.

Summarize Feelings as Well as Ideas

Provide feedback to show you are listening. Acknowledging what others feel shows them that you get where they are coming from.

DON'TS

Don't Judge

Listen without judging. Avoid agreeing (which can alienate others who disagree or fixate people on the problem). Also avoid disagreeing, correcting, and even countering with facts.

Avoid Probing

Questions should be open-ended and clarifying, such as "Can you elaborate some on...?" Speakers should not feel like they are on trial.

Don't Patronize

While praising may be well-intentioned, it can feel patronizing. Avoid phrases like "You've done a good job of..." and "That's absolutely correct, in fact..."

Sandman goes on to note that it is very useful to validate concerns where possible, and to find things you can agree with. But it is crucial to echo even the things with which you disagree. (Again acknowledging that you heard something does not mean that you agree with it. There are plenty of ways to communicate your disagreements, such as through website question-and-answer documents, newsletters, fact sheets, and opinion pieces).

While active listening requires you to identify what is being felt by speakers, that does not mean that they are comfortable with you calling it out directly. Sometimes, stakeholders want their feelings acknowledged but, because of pride or other factors, they do not want you to implicate them. In these cases, you will need to deflect the emotion while at the same time recognizing it. Table 5.2 provides examples of how to deflect emotions.

TABLE 5.2

Examples of Recognizing Stakeholders Feelings While at the Same Time Deflecting Them

Degree of Deflection	Points at	Example
Undeflected	Stakeholder	"You're not really worried about health! You're afraid your property values might be affected."
Deflected	You (person responding)	"I was in a situation like this when I lived near an industrial park. What worried me even more than the health effects was the possibility that my property values might be affected."
More deflected	They (someone close to the situation)	"One of your neighbors was talking with me last week about this situation, and the thing that worried him the most was the possibility of an effect on his property values."
Still more deflected	Some people (someone generic)	"Some people in a situation like this would probably be worried about their property values."
Most deflected	It (something out there)	"It's possible there could be some concern about property values here."

Source: Adapted from *Empathic Communication in High-Stress Situations* (Sandman, June 2010) with permission (https://www.psandman.com/col/empathy2.htm).

5.5 Cool the Water before Wading In

When people are faced with highly technical, complex, or unfamiliar issues that have the power to change their lives, or even the perceived power to change their lives, it can create situations that arouse a lot of emotion or anger, including:

- Anxiety
- Fear
- Defensiveness
- Frustration
- Lack of Control
- Anger

Entering emotionally charged situations is often a concern for those conveying environmental and health risk information. Risk communication expert Vincent Covello teaches that when people are upset, they want to know that you care before they care about what you know. Thus, it is best to allow people to vent their frustrations before moving on. For instance, if an unexpectedly high number of angry participants show up to an informational meeting, you may need to delay or even reschedule some portions of the sessions to accommodate more time for listening up front.

When faced with emotionally charged interactions:

- Let people vent.
- Do not interrupt, be defensive, or argue.
- Respect their opinion and their right to hold it.
- Try not to take their anger or emotion personally.
- Use active listening skills.
- Ask questions to clarify the source of their concern, anxiety, fear, or anger.
- Summarize what a person feels and what you heard.
- Get their agreement on your summary.
- Ask what they would like done to address their concerns.
- If you can do what they requested, then agree and establish a time when it will be done. If you cannot do this, offer to take the request to those who may be able to do it.
- Commit to a schedule and follow up.

Once you have listened, it is critical that you demonstrate that you heard and acted on what you learned. This is true even if your action was to ultimately dismiss a request or disprove a claim. Finally, ensure that your stakeholders know that you followed through; otherwise, it will have been for naught. Whether through direct phone calls, letters, emails, texts, public meetings, websites, or newsletter postings, take all possible opportunities to let stakeholders know you followed through with your commitments.

Appendix A contains a listing of outrage reducers created by Peter Sandman and available at https://www.psandman.com/col/laundry.htm.

CASE IN POINT

Let the Audience Determine When it is Time for You to Talk Shop

At a public meeting on plans to decommission a former nuclear materials handling facility, third-party technical oversight team members who were waiting to speak were instead sidelined while attendees (mostly nearby residents) shouted and broke into chants of "We want sampling now." The verbal mayhem continued for 45 minutes until one resident shouted "Why don't you shut up and let's hear what these people have to say?" Other residents agreed, and the remainder of the two-hour meeting was much more civil and productive.

While it may have been tempting to pack up or push an agenda, the speakers did the right thing in letting the residents decide when they were ready to listen. By demonstrating patience, the speakers also allowed other residents to help maintain order in the audience. (While some measures may have been taken to limit the length of the disruption, the scenario provides an excellent example on the importance of letting stakeholders decide when they are ready to listen.)

References

Adams, S. 2019. *Loserthink: How Untrained Brains are Ruining America*. New York. Portfolio/Penguin.

Carnegie, D.D. 1981. How to Win Friends and Influence People. (Revised Edition.) New York. Simon & Schuster.

Covello, 2011. Basic, Intermediate, and Advanced Risk Communication Skills: Diverse Audience Applications. EPA Record Collections. https://semspub.epa.gov/work/HQ/174861.pdf (accessed March 29, 2020).

Covey, Stephen. 1989. *The 7 Habits of Highly Effective People*. (Revised Edition.) New York. Simon & Schuster Fireside.

Davis, N. 2009. Insight: Sound Science and Effective Communication. Independent Commodity Intelligence Services. https://www.icis.com/explore/resources/news/2009/09/18/9248796/insight-sound-science-and-effective-communication/ (accessed March 20, 2020).

Garber, P. 2008. *50 Communications Activities, Icebreakers, and Exercises.* Amherst. HRD Press, Inc. http://plankcenter.ua.edu/wp-content/uploads/2018/08/50CommunicationActivitiesIcebreakersandExercises.pdf (accessed July 19, 2020).

Kaplan, J., Gimbel, S. & Harris, S. Neural correlates of maintaining one's political beliefs in the face of counterevidence. *Sci Rep* 6, 39589. 2016. https://doi.org/10.1038/srep39589

King & Spalding LLP, 2015. Actual Data, Not Estimates, are Needed for Informed Statements About Risk Associated With Radon in Natural Gas, *Lexology.* https://www.lexology.com/library/detail.aspx?g=a8147269-8583-4e70-848c-f92397c2d759 (accessed August 5, 2020).

Morgan, M.G., B. Fischhoff, A. Bostrom, C. Atman, 2002. *Risk Communication: A Mental Models Approach*. New York. Cambridge University Press.

Patterson, K., J. Grenny, R. McMillan, and Al Switzler. 2012. *Crucial Conversations: Tools for Talking When the Stakes Are High.* (Revised Edition) McGraw Hill.

Pennsylvania Department of Environmental Protection. 2016. Technologically Enhanced Naturally Occurring Radioactive Materials (TENORM) StudyReport. http://www.depgreenport.state.pa.us/elibrary/GetDocument?docId=5815&DocName=01%20pennsylvania%20department%20of%20environmental%20protection%20tenorm%20study%20report%20rev%201.pdf%20

Pennsylvania Department of Environmental Protection. n.d. Center Valley/Coorpersburg Area Radon Response. https://www.dep.pa.gov/About/Regional/Northeast-Regional-Office/Pages/Area-Radon-Response.aspx

Rathi, A. 2016. A Cambridge Professor on How to Stop Being so Easily Manipulated by Misleading Statistics. Quartz. https://qz.com/643234/cambridge-professor-on-how-to-stop-being-so-easily-manipulated-by-misleading-statistics/

Sandman, P. 2006. Risk Perception, Risk Communication, and Risk Reporting: The Role of Each in Pandemic Preparedness. Presented at "The Next Big Health Crisis – and How To Cover It" (later renamed "Avian Flu, a Pandemic & the Role of Journalists"), Nieman Foundation for Journalism, Harvard University, Cambridge MA, December 1, 2006. https://www.psandman.com/articles/nieman1.htm (accessed March 9, 2020).

Sandman, P. 2010. (September 16-17.) Risk = Hazard + Outrage. Excerpt 1. Presented to the RioTintoMiningCompany,Brisbane,Australia.https://www.youtube.com/watch?v=WU__jJzr_Hw, linked from https://www.psandman.com (accessed March 9, 2020).

Sandman. 2016. Confirmation Bias (Part One): How to Counter Your Audience's Pre-Existing Beliefs. http://www.psandman.com/col/confirmation-1.htm (accessed March 9, 2020).

Shank, S. Sleight of Unclean Hand(s): The Cape Wind Controversy, University of Denver. https://www.law.du.edu/forms/writing-competitions/documents/winners/19.pdf (accessed July 19, 2020).

Schmidt, S. 2019. Battlelines Drawn in Fight over Landfill Expansion. Delaware Public Media (June 21). https://www.delawarepublic.org/post/battlelines-drawn-fight-over-landfill-expansion (accessed July 19, 2020).

Song, L. 2014. Leaked Transcript Gives Oil Lobbyist Taste of His Own Medicine. Inside Climate News. https://insideclimatenews.org/news/20141104/leaked-transcript-gives-oil-lobbyist-taste-his-own-medicine.

Upper Saucon Township. 2016. DEP Finds Record Radon Level in Upper Saucon Home, Encourages All Pennsylvanians to Test for this Radioactive Gas. https://www.uppersaucon.org/2016/11/18/dep-finds-record-radon-level-upper-saucon-home-encourages-pennsylvania-residents-test-radioactive-gas/

Ury, W. 1991. *Getting Past No: Negotiating Your Way From Confrontation to Cooperation.* (Revised Edition.) New York: Bantam Books.

Willink, J. 2020. *Leadership strategy and tactics: Field manual.* New York: St. Martin's Press.

6

Don't Skimp on the Public Participation Process

Introduction

If your operation involves risk to the environment or human health, it is not a matter of *if*, but *when*, the public will weigh-in on an important permit application or request for approval through a mandated public participation process. The nature of these mandated public participation processes has matured and expanded over the last 50-plus years. In the mid-1960s environmental interest groups broke new ground with lawsuits that gave them standing to challenge permits. By the 1980s, raucous public meetings became a catalyst for the field of environmental risk communication. Most recently, advances in en masse video conferencing spurred by the COVID-19 pandemic have ushered in virtual participation, including public meetings and hearings.

Regardless of the regulatory trigger for a public participation process or the forum in which it will be conducted, your success will depend in large part on pre-established relationships and shared knowledge among those involved. In addition, to the extent that you can be flexible, inclusive, and open to opportunities of mutual benefit, the greater your potential for positive outcomes.

Your success in conducting public participation will also require you to tend to the *process* as much as to the *content* of the technical material you present. Risk communicators often focus on the content of the technical material at the expense of the communication process itself. Certainly your content must be technically correct and tailored, but if you do not effectively engage your audience, you may be wasting your time.

6.1 Evolution of Environmental Public Participation Mandates

The roots of modern-day public participation in environmental regulatory matters may be traced back to a landmark ruling in 1965 by the U.S. Second Court of Appeals. In this case, a group of citizens, the Scenic Hudson Preservation

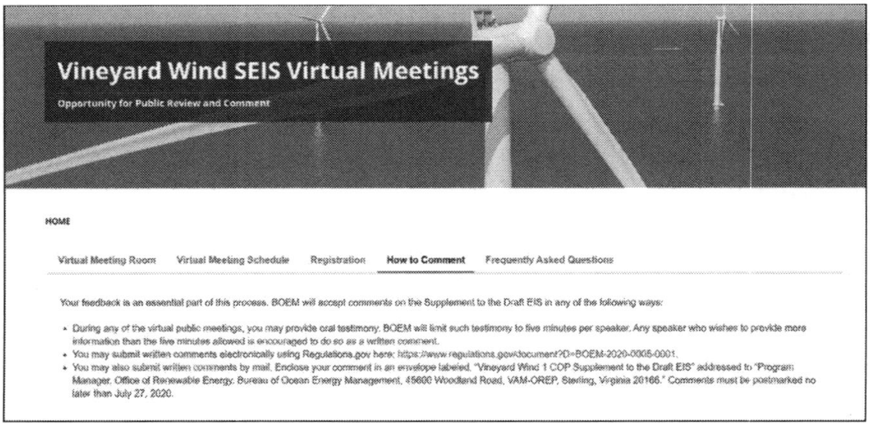

Vineyard Wind Public Comment. Oral testimony on the Vineyard Wind project planned for the coast of Massachusetts was provided by virtual public meeting during the COVID-19 pandemic in 2020. The Vineyard Wind project comes on the heels of the proposed Cape Wind offshore wind project that was ultimately dropped in 2017.

Conference, sued the Federal Power Commission over its approval of plans for Consolidated Edison to construct a power plant on Storm King Mountain in New York. The court determined that the interest group had standing, and thus the case established a precedent for allowing public-based environmental groups to engage in legal processes (Spyke, 1999).

As the Consolidated Edison case played out in court, industries across the nation were drawing nearer to the environmental revolution. The days were numbered when plant managers touted the success of their work-horse industries in terms of spent resources—tons of this and tons of that per day. Local laws to curb pollution in heavily impacted areas, such as early particulate matter control regulation by health departments in Allegheny County (Pittsburgh) and Los Angeles County, began to grow teeth. On the federal level, major legislation such as the Clean Air Act and Clean Water Act were being drafted with built-in citizen suit provisions.

At this developing stage of the environmental movement, written opinion from the 1965 decision on the Consolidated Edison case provided a glimpse of the shortcoming that would plague early public participation mandates: transparency does not equal participation. Wrote Judge Irving R. Kaufman:

> [I]f I were an environmental activist, I would not have any great feeling of satisfaction that the procedures leading to the final decision permitted or, perhaps more importantly, encouraged maximum input and participation by interested and affected groups
>
> … I fear that public participation was far from full or effective in any sense that looks beyond the boundaries of technical openness…
>
> Spyke, 1999

Just more than a decade following the Consolidated Edison Case came three environmental disasters that bolstered citizen's interest in having a hand in the polices that affect them: Love Canal (1978), Three-Mile Island (1979), and the Union Carbide gas tragedy in Bhopal, India (1984). These incidents helped to shape the environmental public participation and right-to-know laws that are the foundation of today's mandates. (Appendix B contains an archive of the Three-Mile Island case study that was included in the first edition of *Environmental Risk Communication*. The partial meltdown that occurred at Three Mile Island remains the worst nuclear accident in U.S. history.)

In the years following these incidents, federal environmental programs began rolling out regulations, policies, and guidance to help agencies and regulated entities navigate the new requirements. Table 6.1 provides a summary of key EPA regulations that provide public participation provisions.

Because the public participation (also referred to as public involvement) requirements were new, the language of many early guidance documents was sometimes disconnected from the reality on the ground. For instance, the introduction to the U.S. EPA Resource, Recovery, and Conservation Act (RCRA) Public Involvement Manual printed in September 1993 states: "Part of your responsibility to implementing any of these activities is to allow those who are interested in or affected by a decision to have the opportunity

TABLE 6.1

Examples of EPA Regulations Containing Public Participation Provisions

Code of Federal Regulations Citation	Topic of Regulation
40 CFR Part 2	Freedom of Information Act (FOIA)
40 CFR Part 6	Procedures for Implementing the Requirements of the Council on Environmental Quality of the National Environmental Policy Act
40 CFR Part 25	Public Participation in Programs under the Resource Conservation and Recovery Act, the Safe Drinking Water Act, and the Clean Water Act
40 CFR Part 51	Requirements for Preparation, Adoption, and Submittal of Implementation Plans (under the Clean Air Act)
40 CFR Part 124	Procedures for Decision-making (EPA procedures for issuing, modifying, revoking and reissuing or terminating all RCRA, UIC, PSD, and NPDES permits)
40 CFR Part 154	Special Review Procedures (procedures to assist the Agency in determining whether to initiate procedures to cancel, deny or reclassify registration of a pesticide product because uses of that product may cause unreasonable adverse effects on the environment)
40 CFR Part 164	Rules of Practice Governing Hearings, Under the Federal Insecticide, Fungicide, and Rodenticide Act, arising from Refusals to Register, Cancellations of Registrations, Changes of Classifications, Suspensions of Registrations and other Hearings Called Pursuant to Section 6 of the Act
40 CFR Part 173	Procedures Governing the Rescission of State Primary Enforcement Responsibility for Pesticide Use Violations
40 CFR Part 271	Requirements for Authorization of State Hazardous Waste Programs
40 CFR Part 300	National Oil and Hazardous Substances Pollution Contingency Plan, Subpart E - Hazardous Substance Response (establishes methods and criteria for determining the appropriate extent of response authorized by CERCLA and CWA Section 311(c))
40 CFR Part 300	National Oil and Hazardous Substances Pollution Contingency Plan, Subpart I – Administrative Record for Selection of Response Action
40 CFR Part 790	Procedures Governing Testing Consent Agreements and Test Rules (Procedures for gathering information, conducting negotiations, and developing and implementing test rules or consent agreements on chemical substances under Section 4 of the Toxic Substances Control)

Source: Public Involvement Policy of the US Environmental Protection Agency, May 2003. EPA 233-B-03-002, http://www.epa.gov/policy2003/policy2003.pdf

to participate in the *decision-making* process." (Emphasis added.) In recognizing that public participation rarely involves the collaborative activities necessary to participate in actual *decision-making*, later updates to this manual, triggered by the need for earlier and more robust public involvement, did not include this language.

U.S. EPA's current universal public participation guidance more closely connects the level of public involvement with the potential for stakeholders to influence decisions or actions. The guidance makes reference to the International Association of Public Participation's *Public Participation Spectrum*, which lays out five levels of participation. The levels of participation range from *no influence*, where public stakeholders are merely being informed of a project, possibly with one or two opportunities to comment, to *total influence*, where public stakeholders are empowered to help make decisions.

When organizations practice "no-influence" public participation, this process is often referred to as the Decide, Announce, and Defend (DAD) approach. While today's public participation processes may be dressed up and tweaked a bit, many continue to reflect the DAD approach to engaging stakeholders. Figure 6.1, adapted from U.S. EPA's Public Participation Guide (undated), provides an overview of the spectrum of public participation. As discussed in the figure caption, the first level of involvement, which involves merely informing stakeholders, represents the majority of public involvement in permitting and approval efforts.

As shown in Figure 6.2, regulatory mandates make up only part of the factors that should come into play when determining the right level of public involvement in a project. For instance, communication plans for sites with low public interest would likely look very different from those with high public interest. Appendix C provides a guide for determining whether public interest in a facility is likely to be low, medium, or high. While the summary was prepared for RCRA-regulated facility projects, the guide should prove useful for other types of facilities.

Whether your project is off the radar or guaranteed to be an all-out battle with deeply rooted activist groups, regulatory public participation mandates should be check-the-box minimums, rather than the thrust of your communication and involvement plans.

6.2 Meet Both the Letter and Spirit of Your Public Participation Obligations

Public participation requirements for environmental approvals generally vary depending upon the potential for the project to impact human health or the environment. For instance, an administrative modification to an existing air permit may have no public comment component while applying for a permit for a waste treatment facility may involve multiple public comment periods, public hearings, radio and TV advertising, and postings near plant entrances.

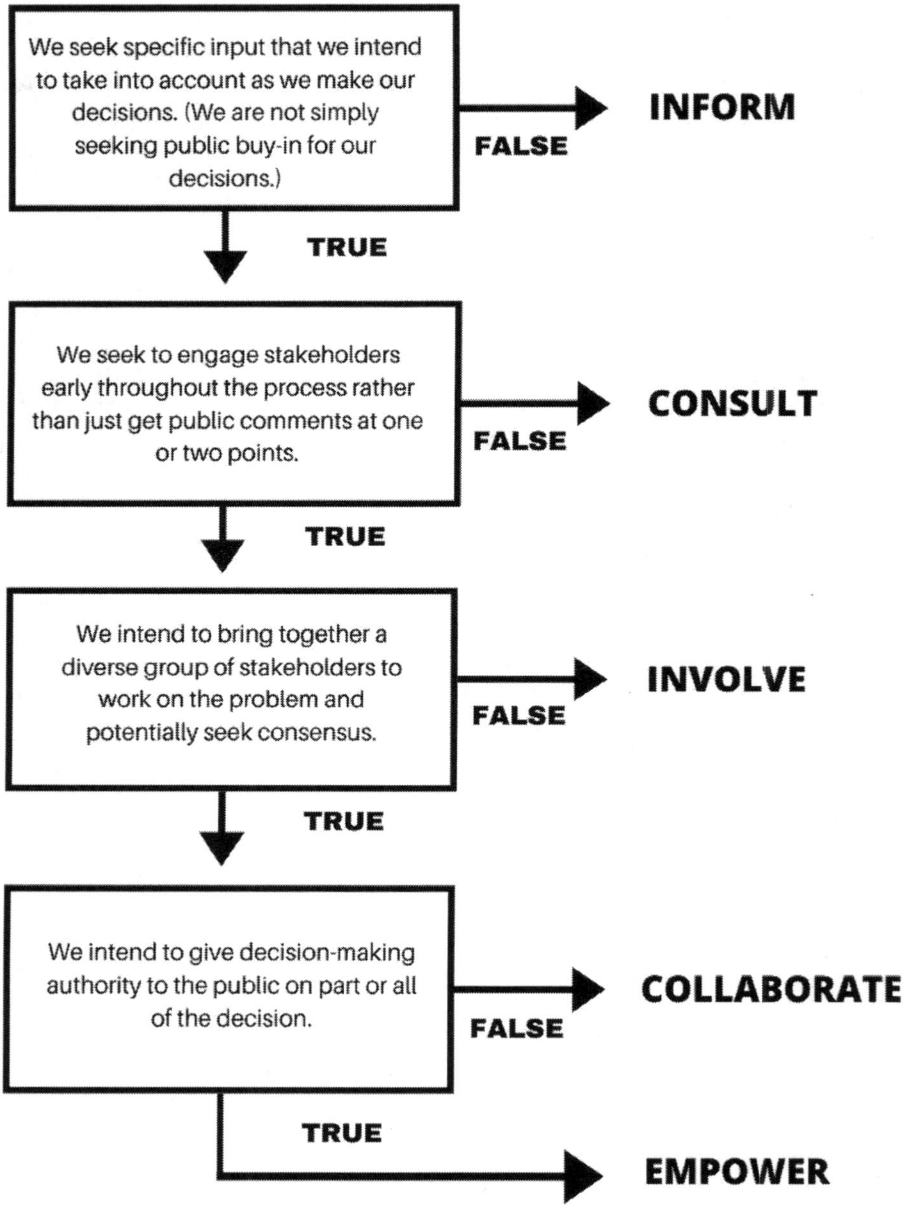

FIGURE 6.1 Public Participation Spectrum. In the public participation spectrum, most permitting and approval efforts do not go far beyond informing the public of plans. (Adapted from *U.S. Environmental Protection Agency Public Participation Guide, undated.* https://www.epa.gov/internationalcooperation/public-participation-guide.)

PUBLIC PARTICIPATION DRIVERS

Statutes and Regulations
- Federal
- State
- Local

Facility Permits, Authorizations and Plans
- Air Quality Permits
- Water Quality Permits
- Waste Management Permits
- Spill and Emergency Plans
- Right-to-Know/Safety Requirements
- Fuel and Chemical Storage Permits
- Zoning and Land-Use Approvals

Corporate Policy and Commitments
- Community Agreements and Pledges
- Corporate/Project Communication Plans
- Industry Initiatives (e.g., Responsible Care®)

Stakeholder Relationships
- Good-will Investments
- Employee Ties to Community
- Advisory Groups

Public Interest and Visibility
- Concerned Citizens
- Agenda-Driven Stakeholders
- Activists and Interest Groups

FIGURE 6.2 Public Participation Drivers. Compliance with public participation regulations should serve as the mere starting point for designing your environmental public participation process. The goals and objectives of the process you design should take into account the bigger picture for your organization and facility.

Giving more weight to public participation. In an unprecedented cause for denial, the Maryland Public Service Commission, in 2006, denied an application for a certificate of public conveyance and necessity by a utility company on the grounds that the company initially had not met expectations for public participation opportunities. Historically, robust public participation had not been required for these certificates. (Photo by Leigh Heasley, Canva.)

If these requirements form the whole of, rather than the bare minimum of, your public involvement effort, you are jeopardizing your chances for a good outcome; you are also jeopardizing opportunities for building relationships that are so important for long-term success.

As a hypothetical example, imagine that you have applied for an approval to perform a dye study of your internal plant drains as part of an engineering project. The aim of the dye study is to uncover any illicit legacy discharges to surface water discharges (a commendable goal). Imagine also that the receiving water for any illicit connections is a stream used by fishing and boating enthusiasts who access the water through a downstream neighboring property owned by a civic club. Because your property spans dozens of acres, the civic club, which generously allows public access and often hosts community fundraisers, is out of sight from your facility.

If, on the one hand, you are an unengaged organization looking to meet the bare minimum needs, you may have no relationship with the civic club and be unaware of the recreational use of the property. You might also feel averse to advertising the dye test because you do not want to raise any unnecessary questions about the integrity of the facility. As a result, you might only pay for a tiny legal notice that can be buried in the classified section of the newspaper, rather than a more expensive display ad that people are more likely to see. Thus, you might end up performing the dye study on a day where the civic club is hosting an "anything floats" fundraiser. Since they were not aware of the dye test, they might think the worst when they see the fluorescent-tinted water heading their way. The result might be a lot of upset people and viral videos on social media. Many of those seeing the viral videos of fluorescent water likely will not see the simple explanation that would follow at some point.

If, on the other hand, you are an engaged community member aware of the club and its recreational activities, you may have talked with them before scheduling your dye test. In fact, you may not have needed to contact them, since you contributed money for the fundraiser and were thus aware of it. Even if these early connections were missed, a more expensive display ad placed in the paper likely would have caught someone's eye and have rung a bell regarding the fundraiser. Since the newspaper reader would know your plant manager from prior outreach efforts, the person might make a call to alert him or her about the event. If your relationship is more established, plant employees may even plan to volunteer or take part in the fundraising, mingling with local residents and answering casual questions about the facility. The result of this outcome would be the continued development of a relationship between the facility and the community, as well as residents who are more knowledgeable about what actually goes on at the plant and the employees who work there.

While hypothetical, these scenarios provide a good example of the differences in outcomes that can occur with an engaged versus unengaged approach to the public involvement process. When weighing the costs of the project investment in the two scenarios, small investments in long-term relationship building and the cost of a larger display ad pale in comparison to the costs of negative press and annoyed, suspicious residents.

As another example, imagine your company is tasked with connecting a resident's conventional gas well to a transmission line. The last area for which you need an easement spans a few hundred feet of unused township property that abuts a neighborhood development. You will have to get the township to agree to the line and negotiate payment for the easement. The extent of mandated public involvement is limited to the township's approval at an open public meeting. The planned line is a four-inch diameter pipe that no one will notice once it is buried. The township will be glad to pocket some extra money. Slam dunk, right?

Not so fast. If the project manager is mindful of the high visibility of the project due to its proximity to homes, and the public sensitivity to gas lines, particularly those associated with fracking (though not the case here, the public is not likely to distinguish), she would do more than fill out the paper work. She might seek out the council member familiar with that section of neighborhood and explain the project. She might also take an hour out of her schedule to meet with that council member and a few neighbors, to show them the area that will be surveyed, and explain the project and the safety measures involved. Perhaps she will present them with a simple overlay map showing other possible routes for the four-inch line that are less desirable. The map might also show other local easements, one of which happens to be for a 20-inch main gas line right behind the neighborhood. She would negotiate a fair and reasonable price for the easement. When the time comes for the township approval at its monthly meeting, she would be in attendance with the overlays. Most in attendance would already be aware of the scope of the project and would not likely have outstanding concerns.

If the project manager were not mindful of sound communication planning, she would just send out the surveyor to determine the most direct path through the park to limit the length of the easement. Curious residents would be eyeballing the surveyor from behind kitchen curtains. Some would talk to him, and would grow concerned because he said something about a new gas line. The project manager might fill out the minimum paperwork, negotiate a bare-minimum payment for right-of-way, and send a junior staff member to the public meeting to button up the approval.

The junior staff member would attend with no handouts and no images to help people visualize what he is talking about. And when people at the meeting bring up concerns about fracking and explosion hazards, he would tell them it has nothing to do with fracking. He might also ask them why they care about a four-inch line but not about the actual high-risk 20-inch gas transmission line that runs behind their properties. He might even grin when they look like deer caught in the headlights because they had no idea they bought properties near a 20-inch gas line. They will feel like chumps and likely demand that the township postpone the vote until they can learn more. And they will likely not have much good will toward the company asking for the approval.

While the facts in each of our gas-line scenarios are the same, the outcomes could be very different. If you managed the planning, relationships, and knowledge-sharing well, it is likely that the gas line would be a non-issue at the public meeting. If you did not manage these issues well, you may be faced with a crowd of people upset about what they think is a major fracking gas line planned for their backyards, and you will spend the meeting trying to dig your way out of a hole. Such an outcome could result in major delays, bad press, a host of higher costs in the form of consulting or legal fees and a higher negotiated settlement.

6.3 Make the Most of Your Process

When outlining a plan of action for your public participation process, think about how the process fits in with your broader communication plans and goals. Doing so allows you to identify opportunities, as well as potential stumbling blocks, for strengthening stakeholder relationships and meeting other long-term objectives. The following are recommendations to help you make the most of your public participation efforts:

- *Set clear expectations*—While *consulting and collaborating* with stakeholders on projects where possible has a greater chance of mutual benefit and relationship building, there is nothing wrong with merely *informing* people—as long as that is what you are calling it. Engagement that is billed as consultation or collaboration when it is actually just sharing of information frustrates stakeholders and makes them lose faith in the process.
- *Look for opportunities to make painless concessions*—Sometimes the little things mean a lot. Watch for opportunities where stakeholders ask something of you that can be easily accommodated even though you are not required to do it. For example, you might be asked to add some people to your mailing list who lie outside the required radius, or extend a buffer of trees beyond the minimum footage.
- *Never go along to get along*—Thinking about the bigger picture means avoiding setting precedents that might be used against you. Something that seems like a painless concession to get a quick approval now could end up being an impediment down the road. This is especially true for negotiations where bad faith actors may be involved and/or project goal posts can be easily moved.
- *Ensure you capture all perspectives*—Resist the urge to only involve stakeholders who agree with you. Do your homework thoroughly enough to ensure that you are capturing stakeholders from all perspectives.
- *Be aware of agenda-driven stakeholders*—Sometimes stakeholders are eager to engage with you out of bad faith. As an example, one company embarking on an effort to be transparent about a controversial operation engaged heavily with the sole person who showed them a lot of interest—a local politician. Unfortunately, this person was on the verge of a coming election and planned to use the information against the company in her campaign.
- *Seek advice from attorneys but keep hold of the reins*—Keeping your eyes on your long-range objectives means you need to weigh legal advice within the context of a bigger picture. In their quest to keep you out of legal hot water, attorneys sometimes give advice that is

unnecessarily restrictive. Where necessary, pressure your attorneys to cut out unnecessary restrictions on stakeholder engagement.

- *Be forthcoming with information*—The more that stakeholders sense you are not sharing information, the more they will insist on finding information and will be suspect of your motives. While most environmental permit and compliance information is available online or through FOIA (Freedom of Information Act) requests, making information easily accessible demonstrates good will. Being able to access information through your website or at a local repository takes pressure off stakeholders who worry that they are being kept in the dark. Chances are they will rarely, if at all, access the information; however, knowing that they can do so at-will provides a real sense of control.
- *Think outside the paradigm*—Limiting yourself to doing things the way they have always been done results in missed opportunities. For instance, financial payments to people directly affected by operations have historically been rejected because of they come across as bribes or payoffs. Section 6.5 discusses the possibility that paying impact or nuisance fees directly to affected individuals could represent a best outcome for all parties.
- *Always close the loop*—Remember to take credit for your hard work of incorporating public comments into your decisions. Do not take it for granted that stakeholders will understand how their input was used in the process. Be sure to explain how input was used along the way. Doing so not only provides you with credit where it is due, it will also help identify and correct gaps between saying and doing.
- *Use feedback to improve*—As mentioned previously in this book, if you operate a stationary facility, you are in it for the long run. Make sure that the feedback you get during the public participation process gets fed into the continuous improvement loop for future communication efforts.
- *Never put all your eggs in one basket*—Having a great relationship with a stakeholder who has tremendous decision-making authority is a terrific asset. But do not count on it always being there. People change jobs, misunderstandings arise, and circumstances change. Think of your outreach network as a spider web—when one strand is broken, the others keep the web intact.

As mentioned in Chapter 3, investing time and resources up front to get the public participation process right is less expensive than attempting a do-over, not only in terms of money and time, but also in terms of good will. Common problems that cause public participation processes to fail include:

- Lack of commitment from within all levels of an organization, particularly from upper management.

- Undefined goals.
- Disconnected functions within the organization, resulting in parallel paths of decision making where public input does not truly influence outcomes.

6.4 Address Environmental Justice Issues

EPA's approach to public participation now includes significant guidance on addressing environmental justice (EJ) concerns. Key federal EJ milestones include EPA's creation of the National Environmental Justice Advisory Council (NEJAC) in 1993 and President Clinton's Executive Order 12898 in 1994. Executive Order 12898, Federal Actions to Address Environmental Justice in Minority Populations and Low-Income Populations, directed each federal agency to develop an agency-wide EJ strategy and to make EJ part of its mission by identifying and addressing, as appropriate, disproportionately high and adverse human health or environmental effects of its programs, policies, and activities on minority populations and low-income populations.

Many states and other non-federal governmental bodies now have EJ programs and policies in place. When embarking on a public participation program, and in communication efforts in general, build EJ concerns into your evaluation. Doing so will help ensure that you are capturing all stakeholder perspectives and any possible additional steps that may be required for public participation. In addition, exploring EJ opportunities might open up possible opportunities for stakeholder grants and partnerships available to impacted communities.

As EJ programs across the country have matured, they continue to face some common barriers to effective community engagement. A federal assessment of progress 20 years after the EJ program was established (NEJAC, 2013) identified the following common challenges to long-term community engagement:

- Availability of resources (specifically, availability of and access to funding and staff to conduct the needed activities over the long term).
- Poor or little coordination among and between various federal, state, and local government agencies and other entities.
- Language and cultural differences.
- Identification of and coalition building among local leadership within a community.
- Lack of cultural competency among agencies trying to cultivate community engagement.

Birthplace of PA Environmental Justice Work Group. The town of Chester, Pa, was the site of an environmental justice battle that led to the creation of an Environmental Justice Work Group under the Pennsylvania Department of Environmental Protection in 1999. A PBS-funded documentary of the case featuring the leader of the grassroots coalition, Zulene Mayfield, may be viewed at https://www.pbs.org/video/justice-in-chester-ajz2de. (Photo by Smallbones, Wikimedia Commons, 2014.)

- Lack of recognition among communities and individuals of their stakeholder status in environmental justice issues.
- Lack of trust between community members, regulatory agencies, and regulated industries.

As many of these issues apply to public participation in general, it is helpful to keep them in mind when developing your process.

6.5 Consider Candid Discussions About Financial Compensation

At first blush, financial payments to community stakeholders impacted by operations may feel like hush money or bribes, but the issue warrants a deeper dive. It is important to recognize that money already changes hands where environmental and public health risks are involved. Fines and penalty assessments are made to regulatory agencies. Host municipality fees are paid to local governments. Impact fees are collected from oil and gas drilling and distributed among government and non-government organizations. Third-party settlements are awarded to organizations such as the Sierra Club and often distributed, at least in part, to other non-profit organizations. Attorneys and consultants are paid substantial sums to support facility siting applications, and so on.

In most cases, this money rarely makes it back into the hands of citizens who are most directly impacted by an operation, such as residents in a host community. As a result, these stakeholders bear a greater portion of the cost/harms, whether real or perceived, than others who are benefiting from those financial arrangements.

As an example, consider the stakeholders affected by fracking operations. In *Fracking Risk Communication* (2013), Sandman points out how fracking often divides a community into three groups:

- *Those who sign leases and collect or hope to collect royalties*—In addition to an increase in regional prosperity (for example, impact fees and employment), this group often receives substantial personal financial benefits from fracking (some become millionaires). They see themselves as beneficiaries of fracking because the potential costs/harms are reasonable given the high benefits.
- *Those who live near drilling operations but have no lease*—This group often receives no compensation outside increased general prosperity in the region, yet they bear substantial cost/harm in terms of noise, dust, truck traffic and other issues. Stakeholders in this group are likely to see themselves as victims of fracking.
- *Those who live further from drilling operations*—This group may realize increased general prosperity without bearing obvious costs/harms. They probably see themselves as more beneficiaries than victims.

This unequal distribution of benefits versus costs/harms comes as a direct result of proximity to drilling operations. Sandman additionally points out that perception of cost/harms can vary as a result of other social factors. For example, unemployed or marginally employed people tend to perceive greater benefits from fracking while wealthy retirees with scenic properties to protect perceive fewer benefits.

Sandman goes on to advise that organizations involved in fracking should look for ways to increase the benefits for those stakeholders for whom the costs/harms are outweighing the benefits. For example, beyond compensating people for specific, unintended harm (like contaminating a water well), companies might consider compensating them simply for putting up with a fracking operation as a close neighbor. Some companies have done this in the form of nuisance payments for truck traffic. (Chapter 7 includes additional discussion of challenges related to the subject of fracking.)

In *Slaying the NIMBY Dragon* (1998), Inhaber suggests that reverse Dutch auctions would allow for more equitable benefits for communities that host NIMBY (not in my backyard) facilities. (An example of a reverse Dutch auction is the way in which airlines ask passengers on overbooked flights to accept compensation for rebooking. The offer would increase in value until a passenger determines the compensation is worth his or her inconvenience and accepts.)

While communities in the United States engage in tax competition for siting new businesses, including some industrial operations, auctions have yet to be used for siting NIMBY facilities. The siting of the Swan Hills comprehensive waste facility in the Canadian province of Alberta in the 1980s remains the only major siting success on record involving community auctions (Mitchell and Carson, undated).

One of the arguments Inhaber made in support of community auctions was that wealthy communities will always escape NIMBY facilities; this observation was true then and remains true today. Take the Cape Wind project, for example. Cape Wind was proposed as the country's first off-shore wind farm project in 2001. The project was to be located some five miles off the coast in Nantucket Sound. Despite overwhelming public support for off-shore wind power projects, including from Massachusetts residents, and initial approvals by the U.S. Army Corps of Engineers and Massachusetts Energy Facility Siting Board, the project was dropped in 2017 after an expensive and protracted battle. Wealthy landowners like U.S. Senator Edward Kennedy and industrialist Bill Koch, whose pristine ocean views were threatened, vehemently opposed the project. While Kennedy and others denied that NIMBY played a role in their opposition, they were widely criticized by environmental groups and project supporters for being disingenuous.

Both Sandman and Inhaber point out that stakeholders who oppose facilities and operations because of imbalances in benefits versus costs/harms are

acting quite rationally. No amount of risk communication aimed at convincing them of environmental and health protections will change that, because the risk is not the issue. Provided that you openly acknowledge the additional burdens faced by stakeholders (not from wrongdoing or noncompliance, of course), chances are that a discussion of direct financial benefits as part of the process might prove mutually beneficial. The closer you can tie financial benefits to the actual stakeholders who are (or perceive to be) bearing the brunt of costs/harms from your operation, the closer you get to balancing the scales in everyone's favor.

6.6 Monitor Citizen Science Developments for Opportunities and Liabilities

Citizen science, that is, the involvement of the public in scientific research or data collection, has proliferated in recent years. EPA describes citizen science as using "the collective strength of communities and the public to identify research questions, collect and analyze data, interpret results, make new discoveries, and develop technologies and applications – all to understand and solve environmental problems." According to EPA, citizen science provides the agency with the following benefits:

- Fills data gaps by gathering crowdsourced data that would be hard to obtain due to time, geographic, or resource constraints.
- Leverages resources by using the efforts of a large group of people to research environmental problems that EPA may not have the resources to pursue.
- Builds meaningful relationships with communities to increase environmental engagement and problem solving and with states and tribes to promote open government.

The promotion and use of citizen science is not limited to the EPA. State and local governments, as well as environmental conservation groups, and activists groups are increasingly engaging citizens in the monitoring of environmental conditions and the collection of data. As an example, watershed groups engage members and volunteers in conducting simple water quality tests. Many state conservation programs, as well as non-profit organizations, engage citizens in monitoring wildlife migration and habitat quality. Some agencies train citizens and equip them to collect air and water samples of sufficient quality to be used as screening tools. For instance, the New York Department of Environmental Conservation developed a Community Air Screen Program that allows citizens to screen areas that may require further study for localized air quality issues. Citizens whose projects are selected are assisted by the Department in identifying the best sampling locations based on their complaints, local meteorological information, and the location of industrial sources or traffic areas. They are then trained to set up and use the equipment. At the end of the monitoring project, citizens receive a report from the agency detailing the results of the sampling and explaining why the assessment will or will not move forward, depending upon the results. In many cases, citizens have found that the source of their complaints is related to pollution from traffic.

Home-based air testing equipment, such as the Purple Air monitor, is a very popular citizen science tool of environmental and health activist groups. This and other next generation monitors have greatly improved in reliability and ease of use. Users can opt to post real-time monitoring results on the Internet (see www.purpleair.com).

While the results of citizen science projects so far have very limited applications to compliance and enforcement, this may change over time. Real-time environmental monitoring equipment will continue to become more sophisticated, reliable, and available. In addition, regulatory standard setting and enforcement are among the spectrum of citizen science uses envisioned by the National Advisory Council for Environmental Policy and Technology (EPA Office of Inspector General, 2018).

Citizen science projects have significant potential for improving transparency and for engaging community members and agencies, as well as for creating more robust databases on industry performance. They also, however, present challenges related to users' lack of understanding of equipment and data limitations, sample integrity, data quality, conflicts of interests, unrealistic expectations for data usage, and varying interpretations by non-expert stakeholders. Thus, citizen science is an area that should be both monitored for potential engagement opportunities and potential conflicts.

CASE IN POINT

Inform Your Process for Best Results

A chemical manufacturer that had historically been a major employer in its host community but had vastly scaled down through the years was embarking on an expansive environmental study of potential contamination from legacy operations. Environmental regulations covering the study and any ensuing cleanup required a minimum level of public participation activities. For instance, the company was required to mail quarterly fact sheets to residents within a certain distance of the facility. Instead of just meeting minimum requirements, the company decided to develop both an environmental project communication plan and a broader communication plan for the operation. In addition to speaking with local opinion leaders, the company also commissioned a local university to perform a telephone survey in the community. To the company's surprise, many residents in the community were not even aware that the facility existed. Thus, educating community members on the facility became one of the goals in the project communication plan. Results of the community research was also used to guide other public participation efforts. For instance, the survey had shown teachers to be among the most trusted people in the community at the time, a fact that helped guide outreach efforts and advisory panel invitations.

References

EPA Office of Inspector General. 2018. EPA Needs a Comprehensive Vision and Strategy for Citizen Science that Aligns with Its Strategic Objectives on Public Participation. Report No. 18-P-02240. Accessed August 2, 2020.https://www.epa.gov/office-inspector-general/report-epa-needs-comprehensive-vision-and-strategy-citizen-science-aligns.

Inhaber, H. 1998. *Slaying the NIMBY Dragon*. Rev. ed. New Brunswick: Transaction Publishers.

Mitchell, R. C., and R. Carson undated. *Protest, Property Rights and Hazardous Waste: A Reassessment.* . https://pdfs.semanticscholar.org/7841/fd123867f01805ccdd1b-7fe7a817ca721887.pdf (accessed August 1, 2020).

National Environmental Justice Advisory Council (NEJAC) (January 25, 2013). Model Guidelines for Public Participation: An Update to the1996 NEJAC Model Plan for Public Participation. https://www.epa.gov/sites/production/files/2015-02/documents/recommendations-model-guide-pp-2013.pdf (accessed July 12, 2020).

Sandman, P. 2013. Fracking Risk Communication. https://www.psandman.com/col/fracking.htm (accessed November 29, 2014).

Spyke, N. 1999. Public Participation in Environmental Decisionmaking at the New Millennium: Structuring New Spheres of Public Influence, Environmental Affairs Law Review. 26(3). http://lawdigitalcommons.bc.edu/ealr/vol26/iss2/2

U.S. Environmental Protection Agency. Public Participation Guide. Undated. https://www.epa.gov/international-cooperation/public-participation-guide (accessed July 12, 2020).

7

Tackle the Tough Issues Head-on

Introduction

To rephrase a John Goddard witticism: *Some companies who find themselves in troubled water wait so long for a ship of good news to arrive on their shores that their pier collapses.* In the meantime, it's too little too late. It's sink or swim, often in shark infested waters.

Difficult challenges in risk communication and public participation may result from any number of situations. In some cases, it is due to the nature of the operation. Landfills, waste treatment facilities, power plants, chemical manufacturers, and other unwelcome but necessary operations suffer from NIMBYism. In other cases, new information may come to light or accidents may occur that cause great concern where none existed before. In still others, your operation may even be the target of activist campaigns.

While no cure-alls exist for dealing with these situations (or any situation for that matter), waiting idly for tough times to pass is not an option. This chapter begins with an exploration of the various factors that may cause

WHATEVER IT IS, WE DON'T WANT IT!

risk communication to be particularly challenging and moves on to discuss approaches and techniques that may be used in these tough cases.

7.1 Factors That Intensify Communication Challenges

As addressed in Chapter 5, there are a number of factors that influence how upset people may become about environmental and health risks. As an example, involuntary risks bother people more than risks they personally choose to accept. Other factors may include level of control, trust, familiarity, immediate versus chronic effects, and transparency. On top of these categories of risk factors, there are other elements that compound communication challenges. Many of these compounding challenges are discussed below

- *Technical complexity*—By their nature, some operations and environmental issues, such as those involving radiation, are very complex and difficult to explain to non-experts. While simplifying these issues for public consumption may require the aid of outside talent, as discussed throughout this book, technical complexity is often not the real obstacle in public communications. A review of the factors affecting risk perception listed in Table 5.1 shows that complex issues trigger many of the "seen as more risky" elements, such as being unfamiliar, able to cause catastrophes, created by humans, and having little perceived personal benefit. Thus, oftentimes, the public may be too outraged to be interested in the details, and technical issues must take a backseat until things cool down. Sections 5.5 and 9.1.2 offer guidance on how to reduce outrage. Section 9.2.1 offers specific guidance on presenting complex technical information in collateral material.
- *Negative campaigns*—Consistent, widespread, and repetitive campaigns by opposition activists can ultimately turn public perception against their targets. Activists often take advantage of the confusion surrounding complex issues by oversimplifying them. While your organization may not perceive their tactics and slogans as a serious threat, they often have real impact on people who know little about the business or have not taken sides politically or ideologically.

In addition to oversimplifying issues, activists often impugn experts who don't agree with them. While this tactic has gained momentum in the current polarized political environment, it is not new. For instance, during a segment titled "Not in My Backyard," on the May 1, 1994 broadcast of the CBS News show *60 Minutes*, the leading local opponent of Von Roll's Waste Technologies Industries (WTI; now Heritage Environmental Services) hazardous waste incinerator stated that "all you need to determine risk is a ruler." The opponent went on to defy any scientist with any integrity to justify the siting of WTI at its present location. (Section 7.6 gives further discussion of anti-industry activists and tactics.)

SOUND-BITE SCIENCE

"...all you need to determine risk is a ruler..."

As discussed in Chapter 5, to counter intense negative campaigns, some industries may need to consider collaborating with trade organizations or other industry partners on counter-offensive campaigns. Better yet, ensure that you and/or your industry partners are communicating about the benefits of your industry before problems begin. And, more importantly, that you are actively *demonstrating* the benefits and reducing the risks.

- *Legitimate concerns*—While industry experts familiar with operations and steeped in regulatory safeguards may have little concern about a facility, nearby residents understandably may dread the plant operations and its emissions, view the plant as an intruder, or may be frustrated by a perceived lack of action or concern on the part of plant owner/operators or the government that regulates the plant (Gerrard, 1994; Trauth & Zuberbuhler Klaber, 1991). It is understandable that the impacted community regards actions that are involuntary, unfamiliar, unilateral, or generally out of their control or participation to be more risky (LeGrega et al., 1994). Sections 7.2 through 7.5 offer guidance for addressing legitimate concerns by demonstrating accountability and providing more opportunities for two-way dialogues.

ON TECHNICAL ISSUES...

Make assumptions and limitations clear and remember that bad data is worse than no data at all. In addition, you must understand and be able to clearly explain any data you publicly present or provide in a permit application.

- *Poor track record*—Some industries carry a heavier burden of legacy issues, whether from histories that predate environmental regulation or wrongdoing by themselves or others. Certainly the public has been negatively impacted relative to property values and health effects by toxic air pollutants, fouled plant water discharges, illegal dumping, and the like. Poor track records should not be left to speak for themselves. Course corrections should be taken where necessary, and should be both demonstrated and communicated.

- *Unfamiliar and emerging technologies*—Unfamiliar technologies that have the potential to be controversial (usually those for which people do not see immediate personal benefits) can be at risk of being framed by those who position themselves as opponents. In a white paper entitled *Communicating Risk in the 21st Century: The Case of Nanotechnology*" (Berube et al., 2010), the authors recommend that advocates of new technologies become better equipped to influence which issues become public by being proactive in public agenda building and helping to shape how the issues are debated in public discourse. They go on to say the following:

> The presence or absence of an issue on the public agenda (priming) and the way that issue is presented (framing) has significant influence on how people think about potential public policy options and form attitudes about technology. Both Greenpeace's 'Frankenfood' frame in Europe and the 'Teach the Controversy' frame are good examples of carefully researched ways of using particular labels or phrases to present an issue that indirectly offer interpretive ways audiences should think and feel about the topic.
>
> Berube et al., 2010

Indeed, taking cues from successful opposition campaigns may have its place in advocacy.

7.2 Being Accountable

As discussed throughout this book, companies should not expect trust and should not ask for it. Trust instead should be earned. However, as risk communication expert Peter Sandman points out in *Fracking Risk Communication* (2013), industries struggling with public trust may be better served by finding ways to be accountable. He cautions that an excessive desire to be trusted leads many companies "... to pass up opportunities to be accountable instead." This is monumentally self-defeating, he says.

Accountability is required by regulators, stock holders, and others directly connected to the business. But, fundamental is accountability to neighbors and the community at the local level. Sandman lists some suggestions for

Site preparation for a fracking operation in northeastern United States Fracking (a term used broadly to capture all operations related to hydraulic fracturing) creates a sizable footprint and a number of temporary nuisance issues in local communities.

local accountability (modified here for applicability to industry in general), as follows:

- "Regularly scheduled neighborhood meetings where concerns can be raised and discussed.
- An 800 number that's staffed 24-7, so a complaint or a worry (or emergency that a neighbor was the first to notice) gets immediate industry attention.
- A full-fledged [Community Advisory Panel, or CAP] whose members can learn the ins and outs of [your operations] and become trusted intermediaries between the company and the rest of the community. (CAPs are discussed in detail in Section 7.3.)
- Treating individual local critics with respect, taking their concerns and recommendations onboard—ideally before they become renowned troublemakers, but at least afterwards.
- 'Technical assistance grants' to local governments or community groups, so the locals can hire their own expert to look over the company's plans, periodically visit [the plant], tell them what to worry about (and what's not really a problem), and pursue concerns with the company as needed.

- A website that automatically and endlessly carries real-time video of [your operations] and key data from monitoring equipment, so a community member who wants to know how things are going can simply check, knowing that if something goes wrong the company can't hide it.
- Negotiations with stakeholder groups, ending in an agreement that at least replicates regulatory requirements and perhaps goes beyond them, in return for those groups' non-opposition—with stipulated penalties (enforceable under contract law without need of recourse to regulatory law) if the company is in violation" (Sandman, 2013).

As Sandman points out, this list contains a range of conventional to daring entries. But its content presents possibilities that industries should consider.

> Of course all the candor in the world won't help until the industry has meaningful things to say about the steps it is taking to reduce environmental risk and the steps regulators are taking to improve oversight. But the reverse is also true: Reduced risk and improved oversight won't be credible—won't even be discussable—until the industry is prepared to acknowledge that the environmental minuses of fracking are real.
>
> *Fracking Risk Communication, Sandman, 2013*

World-wide opposition. Stenciled anti-fracking graffiti in an underpass in Slovenia.

Notwithstanding the actions of industries, regulatory agencies and non-governmental organizations (NGOs) are already putting some of these accountability measures in place. For instance, technical assistance grants are already provided by some federal and state agencies for community oversight or technical evaluation of proposed operations and environmental cleanups.

Grant money given to NGOs is also being used to support real-time video and air-quality and water-quality monitoring networks. In some cases, such networks are being operated solely or in part by regulatory agencies. As examples of this real-time monitoring for public viewing, Carnegie Mellon University (CMU) operates multiple cameras directed at large industrial sites in the Pittsburgh, Pennsylvania area. They also host a mobile application called Smell Pittsburgh (Smell PGH) app. According to their website:

> Smell Pittsburgh is a mobile phone app designed to engage Pittsburgh residents in tracking pollution odors across our region. The app also includes a map-view showing smell reports submitted in the area on a given date. This allows residents to track where odors are frequently concentrated, and link those smell events to poor air quality in, or upwind from, those areas (https://www.cmucreatelab.org/projects/Smell_Pittsburgh).

Organizations whose operations are being monitored by such networks should stay on top of the information available in the public domain and be sure to address legitimate concerns as well as to correct misinformation.

7.3 Establishing Community Advisory Panels

As noted in the *Fracking Risk Communication* tips on finding ways to be accountable (Sandman, 2013), forming a CAP—also known as a community advisory board or citizens advisory committee/group—is a proven method of opening a constructive dialogue with the community that is worth consideration for facilities where community interest and/or concern is high.

A CAP is a group of interested stakeholders formally organized to create a dialogue between a community and a facility and to resolve issues related to the effect plant operations have on the community. An initiative to form CAPs was introduced by the Chemical Manufacturers Association (now the American Chemistry Council) in the early 1990s in response to low public approval ratings of the industry. The CAP format was modeled after citizen advisory committees (CACs) that functioned in a similar capacity for federal agencies engaged in Superfund cleanups. (A number of CACs still operate today.) Since their introduction into the chemical manufacturing industry, CAP models have been adopted by a number of other industries, as have alternative models, such as Good Neighbor Agreements.

GOOD NEIGHBOR AGREEMENTS

"A Good Neighbor Agreement is generally a non-binding agreement between a neighborhood (community) and an industry which works to address specific issues of concern in a collaborative way. It is a way for each group to understand the other's mission and point of view, and results in some form of compromise which can benefit both groups. This approach encourages voluntary actions instead of the more lengthy legal challenges which may otherwise result. It is often best attempted during a permit renewal or some other change at the facility where the company is seeking community approval for their activities."

Good Neighbor Agreements, Eibenholzl, undated. https://doitgreen.org/topics/business/good-neighbor-agreements/

CAPs have been both lauded for fostering community-industry dialogue and criticized by those who suspect them of being little more than public-relations vehicles for the industry. Thus, well designed CAPs must include the two-way flow of information and mechanisms for implementing shared resolutions.

The success and effectiveness of a CAP depends in part on how well its members represent the diverse perspectives of the community. The wrong CAP members are usually "well-intentioned people who complicate things, or people who have an agenda of their own that conflicts with the purpose of the [CAP]" (Farris, 1997). The total number of members typically should be limited to between 15 and 20.

A properly selected CAP can provide you with useful suggestions and insight. The CAP can:

- Act as a liaison between the plant and the community.
- Act as a "sounding board" for new or modified operation changes at the plant that will affect the community.
- Review and provide input into plant operations for safe operating practices and pollution prevention practices.
- Furnish oversight of corrective actions for safety and pollution prevention deficiencies.
- Help to develop effective brochures, newsletters, press releases, announcements, and the like.
- Advise on planning for open-houses, public meetings, and community presentations like those for civic groups and schools.
- Identify sensitive issues and concerns within the community.
- Make suggestions for helpful public service projects that the plant can undertake to enhance the life of the community.

There are many reasons for forming, and some for not forming, a CAP. Perhaps the best time to establish a CAP for an operation that is of high public interest is when community relations do *not* need improving (Farris, 1997).

In forming and operating a CAP, make sure that the committee is not simply a public relations ploy. CAP meetings should discuss and act on issues of substance. The facility must not control the CAP or simply steer it to conclusions believed to be beneficial to the plant. In fact, a neutral, third-party facilitator works best. According to a study of CAPs that were in place in 1999 (Lynn et al., 2000), a professional facilitator was cited as a key factor in CAP success. (Significant commitment by plant managers and diverse membership were other key factors.)

The formation of a CAP should be guided by an experienced professional to help you avoid the pitfalls associated with the endeavor. A CAP is not a good idea if a facility is unwilling to commit to the substantial amount of long-term work required for its continued maintenance. As a short-term alternative to a CAP, a focus group may meet the needs of the facility and community. A focus group is a forum for stakeholders to express their opinions and concerns. Focus groups typically have a neutral, or even sympathetic facilitator who conducts the group sessions at a neutral location.

7.4 Submitting to Independent Reviews

Independent environmental and safety reviews or audits have a place in demonstrating accountability. However, due to a number of concerns, liability chief among them, such reviews would need to be carefully considered and vetted by counsel.

Evaluation of a large, complex operation such as a chemical plant or hazardous waste incinerator can be best accomplished using a team of specialists in diverse areas. A professional review staff should consist of environmental engineers and scientists, environmental auditors, site assessors, field technicians, and experts in pollution prevention, emergency response, health and safety, permitting, compliance, and remedial design (Sadar, 1993). The review staff would be valuable for performing unbiased analyses such as verifying calculations, evaluating assumptions, conducting independent air, water, and soil analysis, and so on.

Such reviews could be a component funded by technical assistance grants through federal and state agencies involved in site permitting and/or cleanup efforts. Independent reviews of proposed siting projects may also fall under this category—for instance technical evaluations of sites that may be considered suitable for construction, as well as the technical design plans for a proposed facility.

Independent reviews were a component in the siting of the Swan Hills hazardous waste treatment facility in the Canadian province of Alberta in the late 1980s, a process in which communities competed through an auction for siting of the facility (see also Section 6.5). The process successfully incorporated the following principles:

- "Need justification.
- Community value judgments of facility impacts and benefits.
- Community choice and decision process" (Zeiss & Paddon, 1992).

This project demonstrated that communities may actually volunteer to host hazardous waste facility (or other NIMBY operations) if they are convinced of the *need* for such an operation, find *value* in hosting the waste facility (traditional siting methods lack this), and are involved in the *decision-making process*.

7.5 Additional Suggestions for Connecting With the Community

All facilities, not only the potentially controversial, should establish as many connections to their host communities as is feasible. Doing so creates opportunities for mutual benefit and helps you get the biggest bang for your buck in terms of outreach and charitable giving. Also, as noted in Section 6.3, community connections may be likened to a spider web—when one strand is broken, the others help keep the web intact.

Below are some additional ideas for outreach:

- *Open houses*—Since *familiarity lessons fear,* the facility may want to hold an open house. Besides giving the public a good opportunity to become familiar with the plant operations, the open house may give the facility a chance to receive feedback on its risk management activities. If open houses aren't an option, you may consider hosting a booth to share information at popular community events.
- *Hotlines*—If your facility is experiencing a temporary nuisance issue, such as odors or dust, or if conditions exist that make it difficult for neighbors to discern the source of a nuisance (for instance between a waste operation and a nearby sewage treatment plant), operating a hotline might be a good idea. If you choose this option, however, you must ensure that calls are answered promptly and must be able to guarantee follow-through. Otherwise, you are likely to make people even more upset with you.

- *Civic involvement*—If possible, encourage employees to become involved in local organizations. If appropriate, you may want to offer incentives, such as time off for volunteering or matching donations. Your employees are your best ambassadors, particularly when they are giving back to the community in which you conduct your operations.
- *Charitable giving*—Every company, large or small, should be engaged in some form of charitable giving. Volunteer hours and/or financial contributions need not be large, but they should be consistent. (Of course, charitable giving should to some extent be proportional to the size and profitability of your organization.) As mentioned in Chapter 3, your organization's principles and goals should help determine how charitable giving is managed.
- *Maintain communications*—Whether you have a communication department that tweets multiple times a day or a skeleton crew that relies on snail mail to connect with the outside world, keep up with the communication channels you have established. If you create a newsletter, don't be overly ambitious in frequency until you are certain you can keep up. If you create a website, don't let it fall out of date.
- *Local government involvement*—If your facility has the potential to cause public concern during permitting efforts (most do), you should be reaching out to local elected officials far in advance to equip them with information about your project. They will probably be the first individuals that residents call with complaints or concerns. (The same holds true for any unplanned incidents that may occur.) Someone from your facility should be attending municipal meetings, including zoning hearings, on a regular basis. Not only does this familiarize you with the local project approval process, it allows you to build relationships with local officials and residents, as well as to be available should people have questions or concerns about your facility.
- *Engagement with first responders*—If your facility must submit right-to-know reports to your local fire department, make sure you offer to have the department tour your facility. By becoming familiar with your operations, first responders are better equipped for addressing on-scene emergencies. If your operations have special hazard considerations, evaluate whether it makes sense to do some joint training exercises. Also, if you have the resources, and it makes sense, you may also wish to have someone from the facility participate in local emergency response authorities that allow for public membership. (Note that whether you actively engage with first responders or not, make sure that you provide them with updates on your emergency plans at least once per year, more often if your contacts change.)

- *Coordination with Public Information Officers/Community Relations Coordinators*—Some local municipalities, regulatory agencies and other groups with whom you may engage have public affairs officials or community relations coordinators. While they cannot work on your behalf or coordinate with you in a way that implies endorsement, they should be updated and kept in the loop on projects of interest to the public.
- *Media relations*—Plan to develop relationships with local reporters who will cover your facility. For large and actively engaged companies, there will be more opportunities and incentives to do this. Media relations are discussed in depth in Chapter 8. As noted there, if you do not have a staff member experienced in media relations, consider getting some outside assistance to help you get started.

7.6 What About Activists?

Earlier sections of this book described several categories of people who will take a position on your facility. Activists generally fall into the category of people who are against your operations. (Note that in the context of this discussion, activists are defined as those who are primarily anti-industry and/or anti-capitalism. Those who simply desire fair and just operating practices, of which there are many groups and individuals, are not being addressed here.)

THE CONTENTIOUS AUDIENCE VIEW

There are no stupid questions…
only stupid answers.

For years, activists have enjoyed being likened to *little Davids* fighting against *Goliath* corporations, defending humanity from hulking, evil industry. However, there are few points of interest here worth noting.

First, in recent years, social media has managed to reverse the perceived balance of power between activists and corporations. Many individuals and corporations once considered untouchable have had their reputations, and sometimes their businesses, unceremoniously crushed by angry tweeters who viciously taunt in anonymity. The viciousness has not been limited to social media. As noted elsewhere, intimidation tactics have escalated from ecotage—a euphemism for vandalism in the name of environmental defense—to forms of personal attacks like doxxing (revealing opponent's addresses and other personal information online).

Second, the real meaning of the David and Goliath encounter—where the giant Goliath falls to the ground in defeat after being struck in the forehead by a stone—actually has nothing to do with size or strength being inherently evil or wrong. Being large and strong does not necessarily mean you are evil, just as being small and weak does not necessarily mean you are righteous. The point of the confrontation in the original biblical teaching is that David had *integrity* while Goliath did not. Ultimately, truth is left standing.

The lesson for industry is that operations must be conducted with integrity. Do the right thing and let those who imagine themselves to be giant-slayers to continue in their vain imaginings. In the end, truth prevails. While good outcomes are not guaranteed, they come more easily when you have done the hard work of walking the talk and building relationships over time.

As noted in Section 5.2, industry has a right, and sometimes even an obligation, to counter mistruths and ensure that factual information and alternative perspectives get floated out into the pool of public knowledge. However, to the extent that activists point out legitimate problems that need serious attention, these issues should be viewed as opportunities for making improvements, rather than a need for more PR. Such PR initiatives can accurately be labeled greenwashing. Greenwashing describes a form of marketing that gives the impression that an organization or its products are environment friendly. Put another way by the authors of *Toxic sludge is good for you!: Lies, damn lies, and the public relations industry*, greenwashing describes "the ways that polluters employ deceptive PR to falsely paint themselves an environmentally responsible public image, while covering up their abuses of the biosphere and public health" (Stauber & Rampton, 1995). Truth be told, both environmental activists and industry share some blame for greenwashing and no one benefits from it. As an example, in *Corporate Green "Disclosures" Are Often Mere Marketing*, columnist Steve Milloy (2020) points out that greenwashed climate-related disclosures in Securities and Exchange Commission filings actually mislead investors and the public because the claims, while they include all the feel-good language demanded from today's companies, aren't tethered to reality.

As a final point regarding dealing with activists, as pointed out in Chapter 4, the extent of activists' influence on your communications effort will be how other groups perceive your treatment of them. You must still share information with them, listen to their concerns, invite them to meetings, and treat them with courtesy. If you don't, other stakeholders may be more sympathetic with them and lose some trust in you.

CASE IN POINT

Bad Actors Contribute to Industry's Reputational Challenges

For years, a foundry coke manufacturer had been ignoring neighbors' complaints about odors, fog, and soot. When citizens complained to their state regulators about the problems, the agency would find nothing wrong based on reviews of emission and permit documents as well as on-site visits with confirmatory sampling.

Eventually, citizens took matters into their own hands by forming a "bucket brigade" of handmade vacuum test kits to sample the air around their homes. After learning that the air had astonishingly high levels of benzene, they shared the data with regulatory authorities. An ensuing year-long investigation by state and federal agencies confirmed that levels of the pollutant exceeded ambient air standards for benzene—by as much as 75 times in some cases.

In the several months that followed the agency investigation, it would be discovered that not only was the facility underestimating emissions, but it was also operating an illegal pollution control bypass that it was able to conceal during agency visits, even when they were unannounced. The illegal activity was only uncovered after a disgruntled employee blew the whistle. In addition to millions of dollars in fines faced by the company, the environmental manager was convicted of 14 violations of federal laws and one count of obstruction of justice. He was fined $20,000 and sentenced to one year in prison and 100 hours of community service.

Adding to the community's anger, a few years later, a charitable foundation headed by the company's CEO elected to donate $1 million each to a college and museum, both miles away from the plant, rather than investing it in the host community or even pollution control upgrades for the plant.

The plant is now closed.

References

Berube, D. M., B. Faber, D. A. Scheufele, C. L. Cummings, G. E. Gardner, K. N. Martin, M. S. Martin, and N. M. Temple. 2010. *White Paper: Communicating risk in the 21st century: The case of nanotechnology*. An Independent Analysis Commissioned by the NNCO (National Nanotechnology Coordination Office). February. Alexandria, VA.

Farris, G. F. 1997. "Citizens Advisory Groups: The Pluses, the Pitfalls and Better Options." *WATER/Engineering & Management* 144 (10): 28.

Gerrard, M. B. 1994. *Whose Backyard, Whose Risk: Fear and Fairness in Toxic and Nuclear Waste Siting*. Rev. ed. Cambridge, MA: The MIT Press.

LeGrega, M. D., P. L. Buckingham, and J. C. Evans 1994. *Hazardous Waste Management*. Rev. ed. New York, NY: McGraw Hill.

Lynn, F., G. Busenberg, and C. Chess 2000. "Chemical Industry's Community Advisory Panels: What Has Been Their Impact?" *Environmental Science & Technology* 34(10): 1881–1886.

Milloy, S. 2020. Corporate Green "Disclosures" Are Often Mere Marketing. *The Wall Street Journal*. May 27, p. A17.

Sadar, A. J. 1993. Assisting the public with their evaluation of proposed hazardous waste incineration, in *Proceedings of World Congress III on Engineering & Environment*. Beijing, China.

Sandman, P. M. 2013. Fracking Risk Communication. https://www.psandman.com/col/fracking.htm (accessed November 29, 2014).

Stauber, J. C., and S. Rampton. 1995. *Toxic Sludge Is Good for You!: Lies, Damn Lies, and the Public Relations Industry*. Rev. ed., Center for Media and Democracy. Monroe, ME: Common Courage Press.

Trauth, J. M., and K. Zuberbuhler Klaber. 1991. *Risk Communication: An Introductory Manual*. Rev. ed. Pittsburgh, PA: Center for Hazardous Materials Research.

Zeiss, C., and B. Paddon. 1992. "Management Principles for Negotiating Waste Facility Siting Agreements." *Journal of the Air & Waste Management Association* 42 (10): 1296.

8

Work with the Media

Introduction

Whether you view working with the media as a tremendous opportunity or a necessary evil, a sound media strategy is a must-have in your communications plan. No other approach in your plan or set of tools in your communications kit will evolve as rapidly as those tied to media communications, thanks especially to social media. Despite threats from shrinking budgets and sinking public trust, the news media will remain a critical stakeholder and public opinion influencer. As with any stakeholder, your aim should be to seek first to understand, and then to communicate.

8.1 The State of Environmental Reporting in America

Journalists of all stripes are caught up in the maelstrom of multiplying media channels, lightning-fast news reach, and boardroom budget slashing. Accuracy and nuance are being crushed under the pressure to be ever faster and more click-worthy.

In 2018, print newspaper for the first time tumbled below social media as a source of news in America (Shearer, 2018). That same year, nearly as many Americans preferred to get news online as opposed to on TV (Pew Research Center, 2019). Such ranking of news sources shift and blend more each year. As Brad Phillips points out in *The Media Training Bible* (2013), the lines that once separated traditional and new-wave media have become blurred as radio and TV newscasters report on viral social media posts and bloggers share their analyses of news from traditional channels: "Today's traditional and new media live side by side, strongly influencing the tone, pace, and content of the other" (Phillips, 2013).

In the increasingly noisy and hectic news environment, not only have audiences' attention spans shortened, they have become increasingly frayed and polarized. Studies by the Pew Research Center and other polling

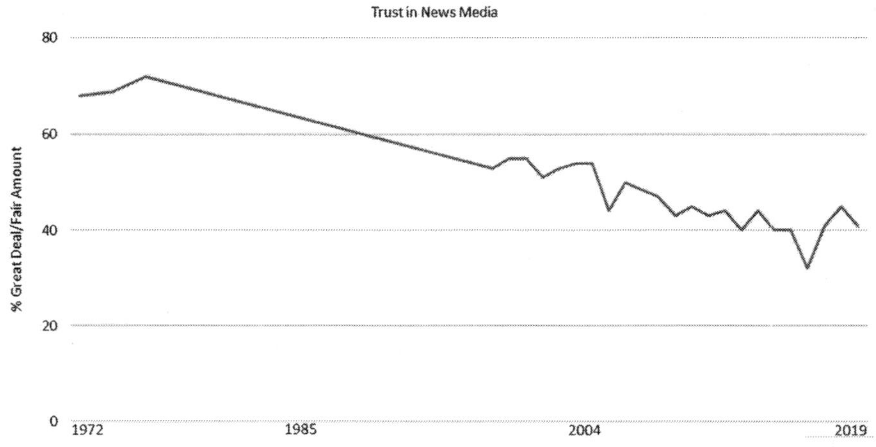

FIGURE 8.1 Public trust in the media continues its overall downward trend. (Data from Gallup, https://news.gallup.com/poll/1663/media-use-evaluation.aspx)

organizations find growing gaps in trust of news agencies based on ideology and political party affiliation—this while public trust in the media continues a long downward trend (Figure 8.1) and as many national reporters who cover polarizing issues, such as the environment and politics, are warming up to the idea of their roles as advocates rather than impartial journalists.

To be sure, today's journalists who report on national and environmental news are increasingly comfortable taking sides. In more than one place, the term for justifying this shows up as "false equivalency." In a *Wall Street Journal* opinion piece on newspaper credibility (Hussman, 2019), the publisher of the Arkansas Democrat Gazette stated this: "Two years ago, I heard a prominent journalist say she doesn't believe in the 'false equivalency' of presenting both sides, and that she sees her job as determining the truth, then sharing it with her audience. That's not what I learned in journalism school in the 1960s."

In a survey released by the Society of Environmental Journalists (SEJ) and the George Mason University Center for Climate Change Communication, the authors state that "A 'false balance' occurs when this approach [giving equal time and weight to both sides of a story] is taken despite a weight of evidence strongly favoring one side over another. In effect, such 'false balance' has the potential to perpetrate an information bias. Scholars suggest that journalistic accounts of human-caused climate change that include an opposing viewpoint are presenting a false balance" (Maibach et al. 2018, page 37). Data compiled from the SEJ survey showed that more than 7 out of 10 member journalists who covered climate change within the prior year rarely or never presented an opposing view point. Only 8% almost always or always presented an opposing viewpoint. This position further shrinks the potential for any depth and nuance of research presentation, which is essential for forming sound policy opinions around such complex issues. Combined with the creep of political bias into news, this pressure on journalists to abandon consideration of opposing views is a powerfully dangerous trend likely to further sow the seeds of distrust and polarization in the public arena.

Locally, where communities have stronger connections to journalists, the news media continues to outperform their national counterparts in the area of public trust. A Knight-Gallup study in 2019 found that 45% of Americans trust reporting by local news organizations "a great deal" or "quite a lot," compared to 31% of national news organizations. Sixty-six percent of respondents said they had more trust in local news versus national news to report without bias (Gallup, Inc., 2019).

The 2019 Gallup poll also reported that environmental issues were among four areas cited as warranting more attention than local news provided. This finding is consistent with results of a study undertaken by Duquesne University more than two decades earlier (MGA Communications, 1995). In the Duquesne study, 56% of U.S. adults surveyed believed there is too little media coverage of environmental issues.

HISTORICAL SNAPSHOT OF STAKEHOLDER PERCEPTION ON ENVIRONMENTAL JOURNALISM

In 1995, Duquesne University in Pittsburgh, PA commissioned a study to determine the need for educational programs for environmental journalists (MGA Communications, 1995).

The study surveyed environmental journalists, corporate communications professionals, and readers.

The target audiences of the study were directly impacted by environmental reporting:

1. *The General Public*—the readers of environmental media coverage.
2. *Environmental Journalists*—the authors of environmental stories.
3. *Corporate Communications Executives*—spokespeople for many of the firms about which environmental stories are written.

Some key results of the study demonstrated perception gaps among these groups:

- 70% of environmental journalists and 56% of U.S. adults surveyed believed there is too little media coverage of environmental issues.
- 60% of environmental journalists said their own news organization was doing either an excellent or good job of reporting environmental issues; however, only 30% of the public and 22% of corporate communications executives agreed.

- Aside from environmental specialists, 57% of environmental journalists felt that reporters were not prepared enough to cover environmental issues, while 74% of corporate communications executives agreed.
- 48% of environmental reporters felt the quality of the environment in America had improved over the last few years, and nearly two in three corporate communicators (63%) agreed. However, only 32% of U.S. adults in the study believed the quality of the environment had improved.

Environmental activists, conservation groups, and outdoor recreation enthusiasts alike use national and local environmental news platforms to share their messages. These groups have adapted well to the changing media landscape, as social media fits nicely with the grassroots campaigning and coalition building that are so important to membership. Information published online by these groups and their members provides a rich repository

The Gold Old Days. Industry experts are no longer the primary source of information used by the news media. Getting your story out now requires consistent messaging through diverse media channels into an increasingly noisy and cluttered information universe. (Courtesy of Turtle Creek Watershed Association, Pittsburgh, Pa.)

of insight into their concerns and interests. Media coverage of stories sourced by environmental groups follow the same trends when comparing national to local news reporting—journalists who are less connected to local communities are more apt to skim the surface and glide in the wake of published stories. And those who have staked out a position are more likely to cherry-pick information.

The current state of affairs in environmental news reporting means a number of things for industry:

- Industries should be placing a premium on relationships with local and regional reporters.
- Interest groups will continue building their networks and use the news "cyclone" to their advantage.
- Locally, people remain interested in environmental issues, even when there are no controversies. Thus, the absence of vocalized stakeholder concerns should not be taken as a lack of interest in environmental issues.
- Companies should be communicating on every media channel that can be justified from a business and outreach standpoint.
- Industry communicators should respect the media as they would any other entity engaged in business trade—fairly, but always with your own interest and objectives in mind. The media have a job to do, and their interests are not your interests (nor are they mutually exclusive).
- All reporters, regardless of their expertise and position, need clear, consistent, and truthful statements in order to perform their job well. These interests are also your interests.
- Misleading information stemming from unsound scientific studies should not go unanswered. Industries impacted by populist reporting should participate in scientific debate, whether individually, through trade associations or alternative venues. This may be in the form of letters to the editor, regulatory comment, support of independent research that meets rigorous scientific standards, participation in advisory boards, or other means.

8.2 The News Business

Journalists look to serve various roles in the mass media, from informing to alerting, educating, advocating, and serving as the gatekeepers of information. A free press and the First Amendment of our Constitution comprise the foundation of our American republic. Beneath its principals and noble achievements over the past two centuries, the press is a business. The business of the press is to sell newspapers, gain online followings, and get high television and radio ratings. The circulation, online following, and TV and radio broadcast ratings determine the fees charged for advertising and that converts into profits for the owners.

To the reporters who work for a news organization, it is a job. They are concerned about advancement, recognition, reputation, job security, earning a competitive salary, and receiving adequate benefits. Like any other workers, they have good days and bad days. They face personal issues at home, they are under constant deadline pressures and they work in an industry that faces the same consolidations, mergers, buyouts, and management changes that are prevalent in other industries.

Journalists hold the First Amendment in very high regard. In all, they strive to do a good job, because their name is attached to every story. They care a great deal about peer recognition, and don't want to look bad within their own profession. Reporters who strive to work in bigger media markets understand that they have to outperform a large number of other reporters with the same ambition.

Corporate communicators must understand what a reporter's job is and what they must do to get it done. All reporters should be treated equally. This includes out-of-town reporters who may come in to cover your facility. (Do not be awed or impressed with a *famous* reporter, but rather treat him or her as a professional, who has a job to do just like you.)

"I JUST DON'T WANT TO DO A GOOD STORY ABOUT A CORPORATION."

Words uttered by a reporter who, upon interviewing an expert about what he thought would be a violation of good environmental practice by a builder, learned that the builder had actually done the environmentally sound thing. The story was no longer newsworthy to him.

Eco-Fads: How the Rise of Trendy Environmentalism is Harming the Environment, Todd Myers, 2011.

While most reporters are not out to intentionally damage a company's reputation, their view of themselves as public advocates is bound to influence their coverage. Nonetheless, the news industry is filled with credible professionals who care about their profession and the impact they have. Likewise, industry is also filled with credible professionals who truly have the best interest of their fellow employees and the residents in mind. Keeping both the people and the business in perspective is helpful in stepping back and reducing the anxiety of dealing with the media.

8.3 What Is News

Every day thousands of air travellers safely reach their destination. This is not news. A plane crash, however, is news. Manufacturing plants safely producing a product is not news. A manufacturing plant that is polluting the air or is the site of an accident is news.

Fires, explosions, chemical leaks, workplace violence—although any of these events may prompt an emergency response from your company, they may not be news. Criteria for selecting what stories make news vary with different outlets. There are, however, some universal characteristics of a news story:

- *Proximity*—What is news in to residents in Newport, Rhode Island may not be news to residents in Newport, Pennsylvania.
- *Consequence*—How many people will your story affect? This characteristic may be the primary consideration of an event at your facility, especially if an off-site release of a chemical is at stake.
- *Prominence*—Individuals cheating on their tax returns may not be news, but if the tax cheat is your CEO, then it will probably be news. When Bill Gates had a pie thrown at him, the event was news because of Gates' prominence in the business community.
- *Novelty*—Building a backyard tree house for your kids is not news. Moving into the tree house to protest the government's policy on timbering is news. Groups will often plan novel media events to gain the attention of the news media.
- *Conflict*—Angry citizens versus a company makes for good ink (or, as noted later, viral capacity). Loud groups of people chanting and protesting makes for good visuals.
- *Human interest*—Nice human interest stories are always in demand by the news media. This is an excellent opportunity for your company to get some positive press by promoting the human side of business. Share your human interest stories with the media as a way to build rapport with editors.
- *Sex*—Consensual sex is usually not news but when the conduct involves someone of prominence it may become news.
- *Digital viral capacity*—In the digital age, some things that aren't newsworthy for traditional media may flourish on social media. This includes the trendy, the innovative, and the sharable.

List based in part on Garvey and Rivers, 1982.

"Journalism is a process in which a reporter uses verification and storytelling to make a subject newsworthy."

American Press Institute, Undated.

If you've been on the front lines at heated public meetings or controversial projects that captured the attention of the news media, you learned firsthand that journalists are in the storytelling business. Oftentimes, it feels like the resulting news stories bear little resemblance to the carefully laid-out facts that were openly shared with reporters. Such was the case during a project to oversee decommissioning of a former nuclear fuels processing plant in Pennsylvania. A six-part evening news in-depth investigation of the project aired after the winter Olympics in 1994 on a Pittsburgh network TV station. Even though the investigation claimed to have carefully examined the history, possible environmental and health impacts, and clean-up activities surrounding the site, the televised report revealed a unilateral story that appeared to fit a presupposed script. Except for a brief camera scan of the masthead from a technical report and mention of the detection of radioactivity off site, results from a 16-month technical oversight effort were largely ignored.

Since it is virtually impossible to determine what events will make news in your area, the best approach is to plan for the worst and hope it never happens. As discussed in Chapter 3, it is important to research and evaluate the possible types of emergencies that may affect your facility. During the research phase of developing a Crisis Communications Plan, a hazards vulnerability analysis should be done to identify possible events that could happen and could ultimately put your company in the news.

8.4 Best Practices for Working With the Media

In a positive outcome from media interaction, reporters get a story with draw, you spread some of your message, and the public receives desired and/or needed information. The potential for overlap of these benefits, however, doesn't mean everyone is working from the same side of the table. Story draw relies on newsworthiness—the sexy, the scandalous, the ire-raising, the inspiring and the innovative—and reporters play up those elements to help sell the news. Your job is to avoid providing fodder while getting out your message.

The principles espoused throughout this book are the same as those best applied in working with the media—act with integrity, be authentic, build relationships in advance, be prepared, and maintain a steadfast focus on your objectives. To do that, mind the following recommendations:

- *Forge productive relationships with reporters*—This means a slow buildup of trust over the long-term. And building up trust means not acting solely in your own interest. Reporters value expertise, dependable information, and valuable connections because these things help them do their job. Find ways that you can be a resource (within the realm of proper professional conduct).
- *Always respond to inquiries from reporters, even if you are prohibited, unwilling, or unable to offer comment*—The *Rule of Thirds* (Phillips, 2013) in media management states that, at best, your message will make up one-third of the story. The other two-thirds are made up of the reporter's voice and your opponent's voice. When you don't engage, your third goes to reporters and opposing voices.
- *Never say "no comment"*—There are a million other ways to say that you cannot engage. For instance, "I can't comment on that until the lawsuit is settled." Simply saying "no comment" gives the impression you are guilty of something or are too arrogant to give the time of day.
- *Respect journalists' needs and constraints*—Reporters work under extreme time pressure, often filing reports from their phones during and immediately after events. If you are unsure of how much time you have to respond to an inquiry or follow up on a discussion, ask. Reporters also take their role as industry watchdogs seriously. Do not

Online News Cycle. Even if your evening public meeting doesn't make TV's 11 o'clock news, on-line reports by journalists are likely to be posted before you pull into your driveway. Established relationships with local reporters and proactive communication beforehand (as well as at the meeting) will increase the chance of more balanced reporting.

expect them to run a draft story by you before printing—sometimes the best you can do is ask for quotes to be read back to you during in-person or phone discussions.

- *Respect the power of the mass media*—As Lundgren and McMakin (2018) point out, the media can have tremendous impact on the financial health of an industry. Consider for example effect of the "pink slime" reporting on finely textured beef in 2012. Beef Products, Inc. was forced to close three of its four U.S. plants and lay off more than 650 workers. Similarly, the Washington state apple industry lost about $130 million in sales in the season following a *60 Minutes* TV program in 1989 about the dangers of the chemical plant growth regulator Alar. Once the media's interpretation of a risk is accepted, perceptions can be extremely hard to alter.
- *Never think you're off the record*—There is no such thing as off the record. Remember this when chatting with the reporter before and after the "official" interview. This applies to non-verbal communication too—any eye-roll, sneer or aggressive pose can be photographed or video recorded for posterity.
- *Don't spin or mislead*—People, and reporters especially, are pretty adept at detecting BS, even when it is not acknowledged out loud. This doesn't mean that you should be voluntarily confessing your sins during your interview, just that you need to find a way to put forth your most positive truths.
- *Do your homework—on both the reporter and the subject.* Every reporter is different, and it is your job to know who you are dealing with before the interview. Review prior stories run by the reporter to get a feel for his or her angle of approach. Know your subject inside out and work through answers to questions before the interview, especially if it is in-person. (Responding to inquiries via email provides for more room here.)
- *Get training if you are inexperienced*—Whether you are extremely confident in your skills or terrified of media interviews, it is wise to seek training before undertaking interviews. Sometimes the most confident public speakers or technical experts give the worst interviews. This is because working through a media interview is not public speaking or grandstanding. Effective interviewees must understand how to get their messages through, even in casual and friendly interactions.

On the flip-side of the training recommendation, being a good spokesperson may not be as difficult as some would think. In *The Media Training Bible: 101*

Things You Absolutely, Positively Need to Know Before Your Next Interview (2013), Brad Phillips, CEO of Throughline Group, notes that inexperienced spokespeople may be surprised that they share some traits with the world's most gifted speakers. According to Phillips, there are six common traits of great spokespeople:

> "*First, they're authentic*—The audience may not agree with their perspectives, but viewers can tell that the spokespersons genuinely believe in their own messages.
>
> *Second, they're natural*—The best spokespersons are the ones the public perceives as being the same person on camera as off, the same in a television studio as in their living room. They're the spokespersons who bring the same passion to their interviews that they express privately when discussing similar topics with their friends.
>
> *Third, they're flexible*—They know that breaking news, technical issues, or a shifting story line can change the nature of their interview with little notice. They know that rolling with the changes, maintaining their composure, and displaying a touch of humor—where appropriate—will enhance the audience's impression of them.
>
> *Fourth, they speak to their audience*—They know that their primary function during an interview isn't to impress their bosses or peers, but to forge a direct connection with each person reading or hearing their words.
>
> *Fifth, they self-edit*—Great media spokespersons know that their job is to reduce information to its most essential parts, never to "dumb down" but always to simplify. They know not to try to say everything, since doing so muddles their message and confuses their audience.
>
> *Sixth, they're compelling*—They know how to express their points in an engaging manner that helps their audience remember them. They know how to use stories, statistics, and sound bites to make their messages stand out, and are adept at coining phrases that stick in the minds of every audience member."
>
> From *The Media Training Bible: 101 Things You Absolutely, Positively Need to Know Before Your Next Interview*, Brad Phillips (2013), CEO, Throughline Group.

8.4.1 Messaging

Section 4.4 explained the importance of messaging, and Table 4.2 provided an example of how to develop key messages. As discussed previously, you should stick to three key messages when preparing to engage with the media. Each of these three messages should in turn be supported with more detailed information. EPA and many others refer to this format of layered messaging as "message mapping."

Messages, particularly those developed for media communication, should be brief—no more than one sentence each. Expert risk communications consultant Vincent Covello recommends that a template of 27/9/3 be used for risk communication messages. This translates into 27 words, 9 seconds, and 3 messages.

Remembering your key messages and supporting points is not enough, however. To prepare for dealing with the media, particularly for interviews, you should repeatedly practice transitioning from potential questions to your key messages—a process called bridging.

8.4.2 Bridging

Bridging, another name for transitioning from a question to the point you would rather discuss, is a critical skill in interviews. In bridging, you briefly answer the question asked by the interviewer and then get back to your key messages.

As indicated in Chapter 3, having a mix of stories, statistics, and sound bites as supporting information helps to reduce the chance you will sound like a broken record when returning to your key messages. Appendix D provides a useful list of bridging statements, courtesy of the Navy and Marine Corps Public Health Center and Fulton Communications.

8.4.3 Guidelines for Successful News Media Interviews

Media interview success is first and foremost the result of effective preparation and repeated practice. Even the most seasoned spokespersons do not skip these steps. In addition to gathering your own messaging and supporting information, assess the audience's information needs and prepare for questions, especially those you hope are never asked.

As mentioned elsewhere, your preparation should not stop at the material you will be addressing. You also want to do some background research on the reporter to learn his or her style and typical approach. Find out what story the reporter is working on, as well as the issues that will be covered. (It is unlikely the reporter will provide you with a complete list of questions beforehand.) Also, try to learn who he or she has spoken with before you, as well as who they plan to speak with after your interview.

Three musts to keep in mind when embarking on media interviews:

- *Get your key messages in early*—You don't know when the interview will be cut off.
- *Stay positive*—Never act defensive. The defensive clips will make it into the news.
- *Remain message oriented, not question oriented*—This requires bridging.

In some cases, reporters may choose to correspond with you via email or text rather than in person or over the phone. While these situations limit relationship building, they offer advantages such as reducing the risk of being misquoted, allowing you to share more background or complex material, and providing a written record of your exchange.

Appendix E contains a laundry list of recommendations for achieving successful media interviews.

8.5 Meeting Media Needs During A Crisis

It is essential that the crisis communication team members understand the needs of the news media. If your crisis communications team members do not have experience in media relations, consider getting some outside help. Firms that offer media training can provide objective critiques of your ability to conduct an interview and your overall preparedness and ability to deal with the news media. This is especially important if your designated spokesperson is a senior level manager and subordinates are reluctant to offer criticism of the boss. Talking to the press during a crisis is very different from speaking before the local Rotary Club.

Knowing what to say and when to say it is a delicate balance during a crisis. Sometimes saying nothing is just as bad as saying the wrong thing. For example, during the Valdez Oil spill, the CEO of Exxon was criticized as being unconcerned due to his failure to go to Alaska immediately after the accident [Fern-Banks, 1996].

Saying the wrong thing certainly will lead to credibility problems. During the Three Mile Island accident, early statements from the utility's management that the incident was under control were countered by federal and state officials saying there is a possibility of a core meltdown (Stephens, 1980).

8.5.1 Who Should Speak?

Senior-level management has to buy into the risk communications process and it is advisable that the CEO be prepared to serve as the ultimate spokesperson for the corporation. If your company employs a communications spokesperson, that professional may be the first to respond to most incidents. If not, trained managers, preferably those who have a connection to the local community, should respond. Spokespeople within the company generally are seen as more trustworthy than hired PR consultants.

If a crisis incident is serious, senior-level managers should be prepared to address the media. This is especially important if there is loss of life. The appearance of the CEO expressing the company's remorse for loss of life demonstrates a high level of interest in the event. The absence of the CEO at the scene may be perceived as a lack of caring on behalf of the company, thus accentuating negative public opinion. (Note that not all situations require that your CEO address the media; in less serious cases, this might be taken as a sign that the accident is more serious than originally thought.)

Assigning the wrong level of spokesperson can be avoided by planning for appropriate company responses to various crises that are identified in your Crisis Communications Plan. A matrix can be developed in your crisis communication plan that matches the type of incident with the appropriate spokesperson. It is important that all spokespersons understand their responsibilities during an emergency and that they have been thoroughly trained in the components of the plan and are comfortable dealing with the media.

All of your employees and contractors should understand how to respond to press inquiries and to whom reporters should be referred. It is not uncommon for reporters to approach receptionists or field employees looking for usable background or quotable information. Employees and contractors should be trained in advance on how to politely and firmly handle these situations.

FROM THE ARCHIVES: UNOFFICIAL SPOKESPEOPLE AND ACCIDENTAL FAKE NEWS

During the 1989 prison riots at the State Correctional Institute in Camp Hill, PA, inmates seized control of the facility and started numerous fires. Because of the security risk, firefighters were unable to immediately enter the prison and extinguish the flames. The local television news media responded, but initially was kept at a distance until the scene could be brought under control.

The television crews set up in the parking lot of a nearby church that was located high on a hill overlooking the prison grounds. This provided a wonderful vantage point for the cameras to clearly pick up the burning cell blocks and the silhouettes of prisoners running loose in the prison yard. The stations broke into network prime time programming and began running the story. Prison officials released information that stated that the perimeter fence was not breached and no prisoners escaped, but the perception from the video was total chaos. It was a classic case where the message did not match the video.

As the situation stabilized, a media area was established near the front gate of the prison, and regular media briefings were held. Despite the frequent briefings, the Camp Hill prison riots provided some valuable lessons for crisis communications teams. Because there was competitive pressure to stay on the air with the story, television stations had difficulty filling time with information. At one point, a volunteer firefighter who got separated from his unit walked around the perimeter of the fence looking for his company. Unable to find his unit, he stopped in the media area and began giving interviews about what he saw. The volunteer firefighter was not a designated spokesperson for either the prison or the emergency management officials and his accounts may not have been factual.

For some unknown reason, the county coroner responded to the scene (even though there were no fatalities during the riots). Another firefighter, apparently exhausted from being on duty for such a long time, decided to climb into the back of the coroner's van and take a nap on the stretcher. He failed to shut the van door and the TV cameras were able to pick up a shot of the coroner's van with a body on the stretcher and fireman's boots dangling from the stretcher. Official word from prison spokespersons was that there were no fatalities, but the reporters insisted that they saw a body in the back of the coroner's van. It was later correctly reported that the body was that of a firefighter taking a nap.

The lessons learned at Camp Hill and other disasters prove that the media constantly needs to be fed information. If the crisis communications team fails to do the feeding, the media will search for its own information, resulting in interviews with lost volunteer firefighters and other *non-expert experts.*

8.5.2 Gathering and Evaluating Information

Providing a clear, accurate, complete, and consistent message is paramount to effective crisis communications. A fact is only as reliable at its source. It is important for the risk communicator to know where to get the needed information as quickly and factually as possible.

The best method to accomplish this is to plan to the maximum extent possible. Company background information, material and waste inventories and descriptions, contact information (including subject matter experts), sample news releases, and other information can be prepared and organized ahead of time so that it is readily available. This information may be stored as bound hard copies as well as in electronic formats for easy access. Equally as important as creating this information is updating it to ensure its accuracy, no less than once per year.

8.5.3 Tips from Behind the Camera

Your communications plan should include a designated gathering place for the media out of harm's way. If the media location is off site, ensure that there is sufficient parking and supplies are on hand to facilitate a news conference.

If a crisis event is prolonged, schedule regular updates to the media. Even if there is little new information released, providing statements will demonstrate that your company is being responsive. Prior to briefings, confirm all your information and be sure to clarify any incorrect information previously released. A member of your crisis communications team should monitor the media coverage and prepare the spokesperson to correct any incorrect information.

During a live television interview, it is important to directly answer the reporter's questions. It is appropriate to use time references during live interviews since the audience is simultaneously seeing the interview. Using phrases such as "What we know now is that there is no release…" or "As of this morning we know…" are appropriate.

For prerecorded interview pieces, avoid time references if possible. Since the interview will be aired at a later time, the past tense is acceptable and recommended. An example, "we had a leak in the container…" "approximately 150 gallons leaked…." Ask the reporter when the interview will air. If information changes after you gave an interview, contact the reporter to ensure that the latest and most accurate information is released.

Television stations may request a spokesperson for a "live interview" during their regularly scheduled newscast. Since many markets have several stations with news shows during the same time slot, careful coordination of your spokesperson will be necessary. Try to accommodate each station without playing favorites.

During the preparation of your crisis communication plan, determine if your local fire department or emergency management agency will co-locate during press briefings. In some states, government officials will not co-locate with the "responsible party" of an accident. If this situation exists at your facility, special coordination and dialog with emergency and government personnel will be important.

In some communities, emergency management officials will establish a Joint Information Center where all the responding agencies and the company will disseminate information. The advantage to Joint Information Centers is that the coordination of information is facilitated and the media has only one place to attend for their reports. Joint Information Centers are widely used in disasters that are caused by an "act of God." If a natural disaster strikes your facility, you will likely see greater cooperation in joint news conferences. If the incident appears to be accidental and the company may be at fault, then government officials will most likely not co-locate for news briefings.

Regardless of the policies in your community, the crucial point to remember is that companies have an ethical obligation to quickly release information if there is a potential danger to employees or residents off site.

A few points to remember when responding to on-site media interviews in front of the cameras:

- *Don't stand in front of your logo*—There is no need to cement in viewers' minds an image connecting the event and your company.

- *Check the background before you set up*—You want to avoid competition from distracting activities, as well as the possibility that controversial or unsafe actions by employees or contractors get caught in the footage.
- *Control the floor*—If you need to keep it short, let reporters know in advance. (For instance, you may tell them that you can take 10 minutes before you have to get back to dealing with the situation and that you will bring them additional information when you have it.)

"HEY, HEY, HEY...WE'RE NOT DONE!"

Words shouted by a reporter who was interviewing the president of Freedom Industries, at the site of a massive storage tank leak into the Elk River in January 2014. The company president, who was reportedly very ill at the time and lacking sleep, flew to the site of the spill to speak to reporters. While his interview was plagued with many missteps, perhaps the most notable was his lack of controlling the floor. After abruptly ending the news conference and walking away from the microphones, he came back to answer more questions after a reporter shouted at him that she wasn't finished.

8.5.4 Speedy Release of Information

In today's environment, neighbors or passers-by with cellphones can stream videos of your disaster to the world in real-time. This and other types of social media feeds have served to further increase the already rapid-response of the media and to provide additional sources of information for journalists' stories. This type of information is ripe for misinterpretation and speculation.

Approach preliminary information cautiously. Confirm all information yourself. Reporters will often dwell on conflicts in information. "Fire officials say this…but company spokesman says …." This type of *he said/she said* reporting may unfairly make your company look bad.

Be prepared to have the live TV crews show up at your facility and be on the air quickly. In the television news industry, there is tremendous competitive pressure to get on the air first with a story. This means that live pictures of your disaster will likely reach residents before you have an opportunity to gather information and release it to the media.

Because of its ability to quickly get the word out, social media can serve as an excellent channel for you to communicate unfiltered information in an emergency. If your company could find itself in the situation where rapid communication of information to the public is critical in emergency response actions, such as sheltering in place, consider establishing social media channels, even if in small measure for now. This will allow your communication to focus on critical public protection actions, even if others are focused on who is to blame or other issues that can be sorted out later.

CASE IN POINT

They Get Their Sound Bites, With or Without You

When widespread environmental contamination from per- and polyfluoroalkyl substances (PFAS) was linked to a prominent chemical manufacturer, the company refrained from addressing the issue publicly, limiting statements to news releases and other controlled channels. After a great amount of pressure, the company engaged directly with the public by hosting a meeting in a nearby community. While the company expressed its desire to be open and transparent, the meeting was tightly controlled. Citizens were only permitted to ask questions in writing, which were grouped for answering. While company representatives and consultants were present to answer questions, they would only permit questions from citizens and news media on prescribed topics.

Understandably, the company wished to control the narrative, prevent the meeting from going sideways, and limit negative press. However, the local TV reporters simply interviewed all of the frustrated attendees outside the building. The resulting news coverage ended up being more about the company's lack of openness and transparency than anything else.

References

American Press Institute. Undated. What makes a good story? https://www.americanpressinstitute.org/journalism-essentials/makes-good-story/ (accessed July 2, 2020).

Fern-Banks, K. 1996. *Crisis Communications: A Casebook Approach*. Yahweh, NJ: Lawrence Erlbaum Associates.

Gallup, Inc. 2019. State of Public Trust in Local News. 2019. https://knightfoundation.org/reports/state-of-public-trust-in-local-news/ (accessed July 2, 2020).

Garvey, D. E., and W. L. Rivers, 1982. *Newswriting for the Electronic Media*. Belmont, CA: Wadsworth Publishing Company.

Hussman, W. 2019. "Impartiality Is the Source of a Newspaper's Credibility," Wall Street Journal Opinion/Commentary, Sept. 10, 2019. https://www.wsj.com/articles/impartiality-is-the-source-of-a-newspapers-credibility-11568109602 (accessed July 2, 2020).

Lundgren, R., and A. McMakin, 2018. *Risk Communication: A Handbook for Communicating Environmental, Safety and Health Risks*, Sixth Edition. Hoboken, NJ: John Wiley & Sons.

Maibach, E., R. Craig, W. Yagatich, J. Murphy, S. Patzer, and K. Timm. 2018. Climate Matters in the Newsroom: Society of Environmental Journalists Member Survey. March. Center for Climate Change Communication, George Mason University. https://doi.org/10.13021/G8S97H (accessed July 5, 2020).

Myers, T. 2011. *Eco-Fads: How the Use of Trendy Environmentalism Is Harming the Environment*. Rev. ed. Seattle, WA: Washington Policy Center.

MGA Communications. 1995. *Environmental Survey*. Denver, CO: Duquesne University.

Pew Research Center, 2019. For Local News, Americans Embrace Digital but Still Want Strong Community Connection. March 26. https://www.journalism.org/2019/03/26/for-local-news-americans-embrace-digital-but-still-want-strong-community-connection/ (accessed July 4, 2020).

Phillips, B. 2013. *The Media Training Bible: 101 Things You Absolutely, Positively Need to Know Before Your Next Interview*. Rev. ed. Washington, DC: SpeakGood Press.

Shearer, E. 2018. Social Media Outpaces Print Newspapers in the U.S. as a News Source. December 10. https://www.pewresearch.org/fact-tank/2018/12/10/social-media-outpaces-print-newspapers-in-the-u-s-as-a-news-source/ (accessed July 5, 2020).

Stephens, M. 1980. *Three Mile Island: The Hour-by-Hour Account of What Really Happened*. Rev. ed. New York: Random House.

9

Rely on Best Practices for Conveying Information

Introduction

As emphasized throughout this book, there is no magic bullet for successful risk communications. Common sense, flexibility, and a focus on your goals will see you through. On the other hand, blind commitment to a particular course of action, and a focus on techniques rather than objectives, will likely lead you astray. There are, however, time- and field-tested principles and practices that you can apply in most situations. Understanding them can save you time and give you confidence in your plans and actions.

9.1 Overarching Best Practices

Under any risk communication circumstances, the following guiding principles, most of which have been discussed at length elsewhere in this book, may be applied:

- *Keep it simple*—Less-but-more-impactful information is the key for both written and verbal communications. Remove ambiguous words and concepts. Get specific by addressing "how" you know what you know, and providing concrete examples. Also, when speaking, know when to stop! Don't keep talking after you have fully responded or made your point.

CLEAR + ACCURATE + CONTEXTUAL + SUCCINT + PROMPT + SENSITIVE + FLEXIBLE COMMUNICATIONS = EFFECTIVE MESSAGING

- *Listen first*—Focus on what your audience wants and needs to know, not just what you feel you need to convey.
- *Always follow through*—Dropping the ball on even small commitments can greatly erode trust.
- *Seek maximum input*—Don't operate in a vacuum. Research your audience and get feedback from others outside your circle.
- *Define and follow your goals*—Know what a successful outcome will look like before you begin. Doing so allows you to focus on the war rather than the battle.
- *Be consistent*—Inconsistencies erode audience confidence. While the level of detail may change with each audience, your key messages and basic information must remain the same throughout your communication project.
- *Recognize and address uncomfortable facts*—Remove attempts to influence opinions by highlighting or omitting information that stakeholders will learn from neutral or adversarial parties.
- *Accept that stakeholder perception equals reality*—Work to inform others' vantage points rather than speaking from your own perspective.
- *Acknowledge uncertainty*—Never discuss risk in a way that implies "zero" risk is possible, because it isn't.
- *Use compatible comparisons and objective terms*—Avoid comparing risks associated with outrage to risks not associated with outrage. Also avoid speaking in subjective terms like "significant," "negligible," and "minor."
- Remember that every word (even those meant to be private but overheard) and every gesture (even subconscious facial expressions) are always on the record and may be memorialized on video.

Other guiding principles may vary depending on the type of risk communication that you are undertaking. The following are some key ideas to keep in mind for each.

9.1.1 Precaution Advocacy Communications

While risk communication often involves dealing with highly upset stakeholders, sometimes the practice requires ratchetting up the concerns of stakeholders who are not worried enough. Such is often the challenge with workplace safety.

Financial impacts continue to highly motivate corporations to communicate and implement effective health and safety policies and procedures. Data from 2018 shows the average cost of a medically consulted injury in the United States was about $41,000. The total annual costs for all injuries and deaths were nearly $171 billion (National Safety Council, undated). And the financial implications of safety go beyond these obvious losses. For instance, companies engaged in construction and consulting may lose jobs to competitors with better safety records. Moreover, poor safety performance and reputations erode employee morale and community acceptance.

These impacts, of course, lie on top of the tragedy associated with loss of life and serious injury. According to the Bureau of Labor Statistics (BLS), National Census of Fatal Occupational Injuries, on average, about 14 workers were fatally injured each day in 2018 in private industry (about 90%) and government service (about 10%). The majority of the 5,250 fatalities involved transportation incidents (2,080). These disastrous events include roadway, nonroadway, air, water, rail fatal occupational injuries, and fatal occupational injuries resulting from being struck by a vehicle. The same census reported that there were 621 fatalities due to exposure to harmful substances or environments and 115 workers died in fires and explosions. But, even more fatalities resulted from "falls, slips, trips" (791) and "contact with objects and equipment" (786) (U.S. Bureau of Labor Statistics).

While great strides have been made in reducing serious worker injuries in the workplace, the pressure for continuous improvement remains.

Like risk communication itself, health and safety programs have evolved over time. Low-hanging fruit was initially captured with changes in policies and training. Continued improvements were then addressed with a focus on employees' attention. Behavior-based systems and peer involved programs like DuPont STOP (Safety Training Observation Program) addressed the issues of continued attention to safe practices and unintended failures to adhere to rules. Employee motivation is now among the next highest level of fruit—figuring out why employees sometimes ignore safety procedures, even when they are properly trained and paying attention.

Literature by risk communication and safety experts that address pathways forward in continued improvements share a number of recommendations:

- *Look for opportunities to use predispositions*—Where possible, build on preexisting opinions, attitudes, values, and expectations, even if they are not related to the safety issue at hand. For instance, if self-expression is highly valued by workers who resist wearing hard hats, perhaps allowing unique (but compliant) hardhats or hardhat stickers to be worn might encourage greater compliance.
- *Seek out, listen to, and where possible, validate, objections from employees*—Is the personal protective equipment ill-fitting? Do employees feel embarrassed wearing it? Maybe these concerns can be addressed without compromising safety. If not, just the act of openly acknowledging concerns may help overcome some resistance.
- *Find ways to provide informal and peer-recognized rewards*—The benefits of rewards over punishments, and of informal and social incentives over formal incentives, have been well-established. Seeking feedback on specific options for rewards can further bolster response.

- *Involve workers in developing and teaching protection plans*—Involved employees are more likely to feel they have a vested interest in the outcome. Moreover, teaching others is proven to be one of the best ways of learning. Involvement should be considered without regard to criticism, as employee outspokenness is often a sign that individuals feel their opinions are being ignored.
- *Stay committed to consistency and repetition*—Consistency at all levels of the organization is imperative to reinforcing safety policies. If employees see supervisors only enforcing rules during visits by corporate heads or regulatory agencies, they will get the message that the rules aren't truly important.
- *Repeat the rules simply and often*—When messages are constantly repeated, people will begin to remember them, just as they do advertising jingles.
- *Boost employees' confidence that their actions will be effective*—Confidence of efficacy is one of the four pillars in the ARCS Model of Motivation (Attention, Relevance, Confidence and Satisfaction) (Keller, 2010). Learners who have no confidence that they have the ability to change outcomes are not motivated to learn. As an example, experiments in prompting homeowners to test for radon showed that communicating the ease of testing for radon significantly affected homeowners' decisions, not only in cases where they were genuinely concerned about radon but also in cases where they were on the fence (Weinstein et al., 1998).
- *Do what you can to make it fun and interesting*—This is tough, but to whatever extent you can arouse curiosity and excitement, the easier to engage others. Digital gamification, the craft of deriving fun and addicting elements found in games and applying them to real-world or productive activities, has potential for upping motivation to the next level.
- *Recognize the rewards that workers reap when they take shortcuts or avoid safety rules.* Each time an employee succeeds in shaving off time, effort, or money by ignoring safety protocols, the behavior is reinforced.
- *Explain how and why safety incidents are reported.* Employees become frustrated when they hear "safety first" but then face push back when reporting an incident. When workers truly believe that production and the bottom line matter most, they are more likely to take shortcuts. Be open about how incidents are reported and what impact they have on the company bottom line; otherwise, employees feel a disconnect that can erode your credibility. Encourage workers to share near-misses as learning opportunities.

- *Offer choices of things to do*—Sandman (2007) suggests that providing a menu of options leads to a deeper commitment to the precautions that employees take. If you surround your list of recommended precautions with two other lists: less difficult ones (what you think is actually the minimum required precautions) and more protective ones (that you think would be over-protective but not problematic), the greater sense of control could help employees comply. This concept might be extended to choices of protective equipment, such as safety eyewear. At the least, a few different models could be offered to allow employees to choose which pair is most comfortable, stylish, and effective for them (Sandman, 2007).

9.1.2 Outrage Management

As discussed in Chapter 5, stakeholders are not ready to listen or engage when they are outraged. Thus, steps must be taken to reduce outrage before communication can progress. Sandman (1998) suggests six principal strategies for reducing outrage:

- *Stake out the middle, not the extreme*—In a fight between "terribly dangerous" and "perfectly safe," audiences will choose to listen to "terribly dangerous." But "modestly dangerous" is a contender. Put another way, if a reasonable observer would rate your effort a B-, activists could get away with giving you an F instead; you, however, will never get away with giving yourself an A.
- *Acknowledge prior misbehavior*—The prerogative of deciding when you can put your mistakes behind you belongs to your stakeholders, not to you. The more often and apologetically you acknowledge the sins of the past, the more quickly others decide it is time to move on.
- *Acknowledge current problems*—Omissions, distortions, and "spin control" can damage your credibility nearly as much as outright lies. The only way to build credibility is to acknowledge problems—before you solve them, before you know if you will be able to solve them—going beyond mere honesty to "transparency."
- *Discuss achievements with humility*—Odds are you resisted change until regulators or activists forced your hand. Now have the grace to say so. Attributing your good behavior to your own natural goodness triggers skepticism; attributing it to pressure greatly increases the likelihood that people will believe you actually did it.
- *Share control and be accountable*—The higher the outrage, the less willing people are to leave the control in your hands. Look for ways to put the control elsewhere (or to show that it is already elsewhere). Let others—regulators, neighbors, activists—keep you honest and certify your good performance.

- *Pay attention to unvoiced concerns and underlying motives*—Unvoiced concerns make the most trouble. Diagnose stakeholder motives other than outrage and hazard, such as ideology, revenge, self-esteem, and greed; your strategy will be different when these elements are at work. If possible, bring them to the surface subtly. Don't waste time trying to change the minds of agenda-driven stakeholders; rather, where possible, address the agendas at the outset. If they are looking for you to buy their property, for instance, the answer is either you can, you cannot, or you will find out and get back to them.

Appendix A provides a laundry list of outrage reducers published by Peter Sandman (2002, reprinted with permission).

9.1.3 Crisis Communication

Be First. Be Right. Be Credible. That is the subtitle of the Centers for Disease Control and Prevention (CDC) 2014 edition of "Crisis and Emergency Risk Communication" and also about the best, most concise advice that can be offered for crisis communication.

In an overview of CDC's crisis and risk communication guide provided to the Allegheny County Health Department Medical Reserve Corps in May 2016, Dr. Elizabeth Felter of the University of Pittsburgh Graduate School of Public Health shared that CDC mantra. In her discussion, she emphasized that one of the core principles of communicating in a crisis is to limit information to the three most important things. The reason for this is that only about 20% of information gets through to people in a disaster because of high mental noise. People take in, process, and act on information differently during a disaster than under normal circumstances.

Other tips shared by Dr. Felter include:

- *Express empathy*—People make up their minds about whether they trust you in mere seconds. And most of what they consider is based on whether they think you care about them.
- *Use positive language*—People may only remember a few words, so instead of "Don't take the elevator," they may only hear "elevator." Therefore, "Take the Stairs," would be a better option.
- *Don't wait for all the facts*—People tend to "anchor" to the first source of information in a crisis. It is OK to tell what you know for certain and what you will share as soon as you know.
- *Plan, test, revise and continue revising plans*—Ninety percent of your communication planning should be done *pre*crisis. Test your plan routinely, and be brutally honest in assessing performance.

- *Don't ignore social media*—Mobile apps and social media combined now make up the main source of people's information in an emergency. At a minimum, post a Frequently Asked Question page—it can help take the place of the bygone phone bank script. (Remember, however, that social media does not reach all populations.)

When it comes to industrial incidents, crises communication efforts are likely to engage attorneys. Such was the case with Freedom Industries when, in January of 2014, a chemical from one of its storage tanks leaked into the Elk River in West Virginia and impacted drinking water in 300,000 homes within a nine-county region (see also Section 3.6.1.). The incident continues to serve as a case study of failures in spill prevention, emergency planning, and crisis communication preparation.

Don Bluedorn, an attorney with Babst Calland, shared his perspective from the firm's involvement in the cleanup and bankruptcy fall-out from the incident during a presentation to the Allegheny Mountain Section of the Air & Waste Management Association in December 2017 (Bluedorn, 2017). These tips are summarized below:

- Don't speculate—Run all the details to ground.
- Establish clear communication protocols at the outset, for both internal and external personnel. Follow them.
- Discuss what is appropriate for phone calls versus emails. Remember that everything in writing could be subject to discovery in post-crisis litigation. (Said Bluedorn, "I would rather get three phone calls in a row than deal with one problematic email or text down the road.")
- Know that agreeing to be a point person translates into a 24/7 commitment and availability.
- Act calmly and deliberately, but without delay.
- Be empathetic when interacting with third-party stakeholders. You will only get one chance with them (if you are lucky).
- Err on the side of over-communication.
- Be accurate with communication. State what you know and what information is still being sought. Know when to say "I don't know."
- Retain a sense of humor—you are going to need it!

Not all crises will be due to technical error or natural disasters. Some may come as reputational attacks by competitors and opportunists. In these cases, it often doesn't behoove a company to follow the average playbook. *In Damage Control: The Essential Lessons of Crisis Management,* Eric Dezenhall discusses that in these situations in particular, it is important to know when to play offense and to understand why and when to fight. His book (Dezenhall & Weber, 2011) offers nine key features of companies (and individuals) best equipped to survive crises, as excerpted below:

- *They have strong leaders* who have broad authority to make decisions.
- *They question conventional PR wisdom* and do not worship at the altar of feel-good gurus who espouse "reputation management," the canard that corporate redemption follows popularity.
- *They are flexible,* changing in course when the operating climate shifts (which it usually does).
- *They commit significant resources* to the resolution of a crisis with absolutely no guarantee that these resources will provide results.
- *They have a high threshold for pain,* recognizing that things may get worse before they get better.
- They think in terms of baby steps, not grandiose gestures, which explains Rome's success, after all.
- *They know themselves,* and are honest about what kinds of actions their culture can—and cannot—sustain.
- *They believe that corporate defense is an exercise in moral authority,* and that their critics are not necessarily virtuous simply because they purport to be standing up for the "little guy."
- *They are lucky,* often catching unexpected breaks delivered by God, Nature, Fortune, or some other independent factor.

Effective management of crises situations requires that good leaders be nimble, willing to do what is best even when it is difficult, and listen to your gut instincts, despite the background noise.

9.2 Advice Based on Method of Communication

Invariably, your risk communication project will entail sharing information through different methods and under different circumstances. For ease of reference, recommendations are provided below for common applications. The advice represents wisdom gained from both personal experience and published recommendations by academic heads and other seasoned professionals in the field of risk communication.

9.2.1 Visual Aids and Print Materials

Whether our attention spans have shrunk below that of goldfish, as claimed by some research interpretations (McSpadden, 2015), or we are simply losing interest in topics more quickly as other research claims (Technical University of Denmark, 2019), readers of visual aids and print materials are scanning more quickly than ever. That means you have to do more with less. Here are some time-tested ways to do that:

- Keep it simple and short.
- Stick to one point per graphic.
- Include your conclusion on the graphic itself.
- Use sequential, or step-through graphics to simplify complex information.
- Use bar graphs and pie charts where possible. If scatter plots are used, explain their meaning carefully.
- Use color to convey meaning, but remember some people are color blind and pieces may be reprinted in black and white.
- For numerical information, Lundgren and McMakin (2018), provide the following tips:
  - Highlight the most important information.
  - Pretest symbols and graphics.
  - Align data with general thinking (for example, in a choice of one to five, the highest number would be best).
  - Give visual clues as to the importance of information (for example, use large fonts or bold items).
  - Consider expressing risks as absolute risks (1 in 10) as opposed to a relative risk (10%), and do not use decimals.
- Avoid acronyms and jargon.
- When discussing risk:
  - Compare risk assessment results to standards set by credible, third-party authorities, such as the EPA and the Agency for Toxic Substances and Disease Registry.
  - Address uncertainties in risk assessment by giving examples, for instance, how an assessment of risk from water contamination assumes that everyone at all ages and health conditions is drinking a lot of water over many years.

Again, feedback from objective readers who best represent your audience is your very best guide to producing effective visual aids.

Figure 9.1 illustrates simple infographics used to depict action levels for cancer and noncancer risks.

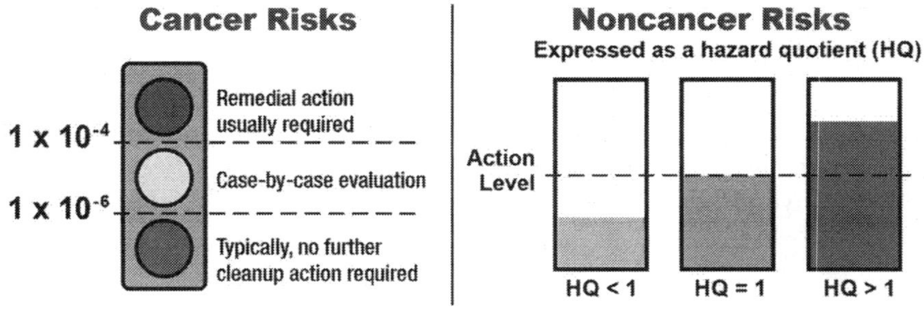

FIGURE 9.1 Infographics illustrating action levels for cancer risks and noncancer risks. (Note: Original infographic printed in color.)

9.2.2 Presentations

Whether you will present to a friendly or hostile audience, there are a number of best practices you can apply in the preparation and delivery of your presentation. One piece of advice that is often left out, however, deserves your attention—be yourself. People appreciate and trust authentic. While you still have to mind all of the potential distractions and follow tricks of the trade, doing so while being yourself can add tremendous value and impact.

Preparation and practice is essential in delivering a quality presentation, regardless of your command of the subject. Even if the engagement is a hurried affair, take time to step back and put your presentation in perspective. Who is your audience? What do you want to achieve with the presentation? What are the key points or questions that need to be addressed?

Being mindful of what type of communication you may be engaged in may also help you prepare. For example, if you are addressing a potentially controversial permitting issue and you expect you may face some initial outrage, be prepared for a Plan B. Maybe you will need to take time to listen to concerns first, possibly putting your intended information on hold for a second meeting.

Always anticipate questions and develop answers (Q&A) ahead of time for sharing among your team. Be absolutely sure to ask yourself the toughest and most uncomfortable questions you might get—these are the most important.

Before going live, practice delivering both your presentation and your Q&A in front of an objective audience willing to give you honest feedback. This is not limited to your client or supervisor—ideally you should also find someone who is as likeminded as the actual intended audience as possible.

Note that your presentation planning stage should also include steps for assessing your space, platform, and logistics. For instance:

- Pretest audio-visual equipment in the same fashion as it will be used together. (Not all equipment is equally compatible.) If your presentation will be hosted in person, test the actual computer, video, and audio equipment that will be used. If your presentation will be done as a webinar, test run each type of delivery feature (e.g., slide show, videos, live frames) with your host and moderator.

- For in-person venues, pack spare equipment parts and extra copies of your presentation. Also assess the room lighting, HVAC, and comfort of audience seating.
- Confirm someone will be on hand to help with technical difficulties.

Appendix F provides handy advice related to planning for and giving presentations. Appendix G provides a laundry list of advice on presentations and handling question-and-answer sessions. Appendix H, excerpted from the Navy and Marine Corps Public Health Center publication *A Risk Communication Primer* (undated, with credit attributed to Fulton Communications), provides tools and starter language for answering tough questions.

9.2.3 In-Person Meetings and Communications

Meeting face-to-face with stakeholders, whether elected officials, regulatory agency staff, advisory panel members, or other stakeholders, can greatly enhance communications and relationship building. This can be especially true during social events and at chance encounters.

Not everyone in your organization, however, will be comfortable or well-suited for face-to-face discussions with stakeholders. Individuals who shy away from social interactions should not be pressured to participate. Similarly, not all individuals comfortable with socializing are necessarily best suited for nurturing trusting relationships.

As noted in Chapter 5, establishing trust is essential to effectively communicating about risk. Outside the realm of risk communication, trustworthiness (also defined as friendliness and warmth), is one of two lenses through which people measure each other up at first sight (Fisk, 2018).

Thus, when choosing who you would like to represent you in stakeholder meetings, ensure that your candidates are individuals who others view as trustworthy. Other tips to keep in mind for face-to-face connections with stakeholders are as follows:

- Ensure that everyone on your team is singing from the same sheet of music. Mixed messages erode trust. To help keep everyone on the same page, update your employees on public outreach efforts and the company's position regarding issues that may attract public attention.
- When setting meetings, involve a proportionate number of representatives (for example, if you are visiting one or two people, do not bring five from your team).
- Be mindful of your listening skills and body language.

- Avoid making any commitments, even very small ones, for which you cannot follow-through with certainty. In fact, where you can, under-promise and over-deliver.
- Mind the dress code. Try to avoid overdressing or underdressing. If you are uncertain, dress slightly up rather than down. Be yourself—don't try to dress exactly like the audience to fit in if doing so will come across as phony.
- Don't badmouth people (employees or outside stakeholders) to others after you leave meetings. Not only does this set a bad example, it prevents team members from trusting you and from working with others in good faith.

9.2.4 Public Meetings and Hearings

Community meetings and hearings are often required public-participation components of larger environmental permitting projects. While public participation and comment provisions of many environmental regulatory programs have expanded in recent years, outreach efforts by the regulated community largely have not developed in equal measure.

For many companies, public meetings and hearings still mark the beginning of stakeholder involvement because they trigger official public comment periods. However, when meaningful stakeholder participation is conducted, all of the heavy lifting is done in advance of the public meeting or hearing. In these cases, misunderstandings and manageable concerns have already been flushed out and resolved.

Fears about opening up to the public any more than is absolutely necessary are understandable. Lots of projects have been derailed or cancelled by vocal and persistent opponents. Laying low can feel like the safe option. However, this approach often relies more on luck than strategy. For stationary facilities that will inevitably require additional permits or approvals down the road, luck eventually runs out. Moreover, choosing to lay low sacrifices up-front opportunities to properly frame discussions and share factual information. When news later hits the fan, as it often does, discussions are controlled by opponents who claim the company tried to sneak a permit through. For these reasons, early and open involvement lays the groundwork for success. And in cases where companies have already established relationships with community stakeholders, early and open involvement is a natural fit.

MEETING DISRUPTERS

If your challenger is taunting you or asking you questions for the sake of provoking or starting a controversy, perhaps to generate media attention, it is critical you do not take the bait. Your job is to remain passive and not to respond or escalate the interruption. Escalating could make a bigger issue out of the interruption than it might be if you remain quiet and calm and do not react. If you or your host organization are aware prior to your talk that there may be interrupters, it is important that you let law enforcement or security know and ask them to be on hand for the event. These incidents are their responsibility, not yours. They will also direct you not to provoke any protestors or respond to their interruptions at all.

Winning Your Audience, Rosebush, 2020.

Thus, the first recommendation regarding public meetings and hearings is to ensure that they fall near the end of your communication and outreach schedule, not at the beginning. Other recommendations are:

- Reference the tips on planning and giving presentations, as well as handling tough questions (Appendices F through H).
- Be patient when attendees resist designated seating arrangements, traffic flow design, or sign-in requests at your venue. Force-fitting a particular format is less important than allowing attendees to participate at their comfort level.
- Expect that you will be video recorded, particularly when activists are involved.
- If you choose to video tape the meeting yourself, whether for your personal record or for assessing your performance, post a sign at all entrances to alert attendees that the meeting may be recorded. For small meetings where attendees have established relationships, mention your intent to record before the meeting begins and seek approval.
- Set expectations for behavior at the outset of the meeting. If physically threatening behavior is a possibility, arrange for security to be present. If you anticipate rude behavior, be clear about what cannot be tolerated and the consequences (closing or rescheduling the meeting). By the same token, avoid using time constraints and rules of engagement as excuses to shut down conversations. For example, instead of cutting the meeting off at a particular time, consider querying attendees if they would like to extend the time. You may also request a separate meeting or side bar for a particular issue serving as a roadblock to the real purpose of the meeting.
- Where appropriate and possible, use a trained, third-party facilitator to keep the meeting on track and enforce rules of engagement.

- Ensure that your designated speakers are individuals whom the attendees find to be the most credible and respected in your organization. Often this is not a person at the presidential or CEO level but rather at the plant manager level.

9.2.5 Social Media Communications

Many established industries, especially the NIMBY-plagued, have waded into the arena of social media with as much enthusiasm as one would a root canal. While waiting on the sidelines, they've witnessed social media tidal waves crush many reputations while buoying others to unexpected heights. Facebook posts and tweets launch into the digital world like pinballs, sometimes silently swallowed into black holes, other times gaining momentum as they light up random targets, and most times falling somewhere in between.

Despite about two decades of use, the market shows no sign of peaking. New features and restrictions change weekly, as does the leaderboard of platforms with the most number of users. The lack of control is uncomfortable, the financial investment is difficult to estimate, and the risks versus benefits are difficult to weigh.

There are, however, a few universal truths that you can count on:

- Social media will continue to grow as a key communication vehicle for reaching stakeholders. In a 2016 study (Gottfried & Shearer, 2016), Pew Research Center found that 62% of adults in the US got their news from social media. Pew Research Center data also show that social media use continues to increase even among the oldest adult populations (Figure 9.2). These trends will undoubtedly continue.
- Social media is a favorite tool in the kit of energized activists. Both well-intentioned conservationists and hostile agitators effectively use Facebook, Twitter and other social media channels to publicize, criticize, and mobilize.
- Social media channels offer unprecedented reach for emergency and crisis communication. Outside of cell phone alerts used by public agencies, there is no faster or more direct line to stakeholders than social media. Day-to-day, social media tools like YouTube provide for one-way, unfiltered data sharing and storytelling that was never before available to corporations. (Unfortunately, this works for misinformation as well.)
- The competitive pressures exerted by social media on journalists to be the fastest and the most provocative will continue to degrade the accuracy and quality of news. Accepting this fact will help industries plan to compensate in other communication arenas.

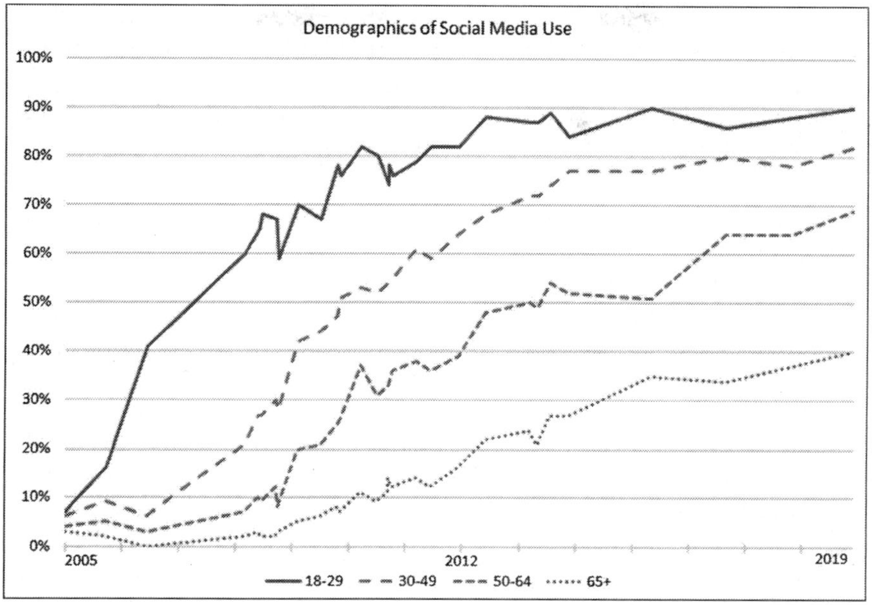

FIGURE 9.2 Demographics of social media use.

- There are no hard and fast rules for which social media platform and which level of engagement is right for any one company. Weighing potential social media options against communication goals is your best bet for employing social media to your advantage.

Given these certainties, there are some actions you can take to assist you in determining the who, what, where, and when of social media:

- Start in the basement. There is one foundational activity for social media that every company should employ—monitoring. Ensure that someone is monitoring social media for references to your company and your business. Also ensure that this information is shared inside the organization with anyone involved in PR and outreach. You can monitor web content for free using no-cost applications like Google Alert, or you can purchase more sophisticated media monitoring services. If you don't know what is being said about you by a substantial part of the community, you have no chance of steering the conversation. Monitoring is especially important for catching misinformation and reaching critical stakeholders in emergencies.
- Stake claim to your name. Even if you choose to keep websites or social media handles dark, or inactive, you should consider registering them before someone else does. This is particularly

true for long-established names, well-known brands, and company names expected to remain unchanged. (For those interested in using Twitter for communicating in emergencies, a low-frequency tweeting strategy, for example maybe once per week or month, might be a better option than waiting to make the handle live.)

- Establish a social media policy for the company and for the employees, and make it well known. The ability to post information on corporate social media networks should be limited to a few trusted employees. (Log-in and password information should be shared among these people, and not kept with one individual.) While you cannot regulate how employees engage in social media outside of work, you can enforce appropriate policies for dos and don'ts through human resources. This will help prevent crises involving reputation and proprietary knowledge by disgruntled, rogue, careless, and even well-meaning employees.
- Prioritize your engagement based on your communication goals. Choose only the platforms and types of social media activity that will help you meet your goals. Recognize that even carefully selected pathways will face challenges and misfires. In addition, your strategy cannot remain static, given the rapid pace of change in the world of social media.
- Ensure that your social media strategy defines how and when you will respond to misinformation. Not every attack or mischaracterization deserves a response. As consistency in response is important, you can't make up your strategy on the fly.
- Understand that more engagement means more internal resources to create, edit, and track content. Given the rapid reaction time in the social media realm, this work cannot be switched on and off or handled as time allows. Also understand that more engagement also means more chances for things to go wrong—more opportunities for campaign high-jacking, misinterpretations, malice, and mischief.
- Remember that everything you publish becomes memorialized. Think twice before pressing the send button and never press the send button while you are upset.
- Also remember that Twitter users do not represent the world. According to the Pew Research Center (Wojcik & Hughes, 2019), most users rarely tweet, but the most prolific 10% create 80% of the tweets from adult U.S. users. While Twitter mobs have enjoyed great power over easily-cowed targets, granting them more power than they hold will undercut you and your partners in industry, especially in the long run.

CASE IN POINT

Smarter than Thee—They Won't Agree

Residents came to a municipal meeting to weigh in on zoning exceptions for a neighbor's proposed gas well. In attendance besides residents were a number of oil and gas industry representatives, their consultants and attorneys, and happy customers.

Among the residents' concerns being addressed was noise. Unbeknownst to attendees, a consultant was measuring noise on a sound-level meter throughout the entire meeting. Just as the discussion on noise became heated, he stood up, displayed the meter, and announced that noise from drilling would be so low, it would register way below the meeting noise he had been measuring.

While the consultant may have proven a factual point, he surely didn't score any with concerned citizens. The tactic felt deceptive and sent the message that the consultant wanted to surprise them about their lack of understanding.

How different would it have felt if he had invited audience members at the outset of the meeting to watch the meter for themselves? In so doing, he may have started the conversation like this…"I understand some of you came tonight because you have concerns about noise…"

References

Bluedorn, D. 2017. Environmental Crisis Management: Lessons from Freedom and Beyond. Presented to the Air & Waste Management Association Allegheny Mountain Section. December 14.

Dezenhall, E., and J. Weber, 2011. *Damage Control: The Essential Lessons of Crisis Management*. West Port: Prospecta Press.

Fisk, S. 2018."Stereotype Content: Warmth and Competence Endure." *Current Directions in Psychological Science*. 27(2) (March/April): 67–73. Article first published online: February 28; Issue published: April 1. https://journals.sagepub.com/doi/10.1177/0963721417738825 (accessed June 26, 2020).

Gottfried, J., and E. Shearer. 2016. News Use Across Social Media Platforms [Internet]. Pew Research Center's Journalism Project. http://www.journalism.org/2016/05/26/news-use-across-social-media-platforms-2016.

Keller, J. 2010. *Motivational Design for Learning and Performance*. Rev. ed. New York, NY: Springer.

Lundgren, R. and A. McMakin, 2018. *Risk Communication: A Handbook for Communicating Environmental, Safety and Health Risks*, Sixth Edition. Hoboken, New Jersey, John Wiley & Sons.

McSpadden, K. 2015. You Now Have a Shorter Attention Span Than a Goldfish, *Time*. https://time.com/3858309/attention-spans-goldfish/ (May 14) (accessed June 9, 2020).

National Safety Council, Covello, Undated. Work Injury Costs. https://injuryfacts.nsc.org/work/costs/work-injury-costs (accessed June 3, 2020).

Rosebush, J. 2020. *Winning Your Audience: Deliver a Message With the Confidence of a President*. Rev. ed. New York, NY: Center Street.

Sandman, P. 1998. Reducing Outrage: Six Principal Strategies. http://www.psandman.com/handouts/sand42.pdf (accessed June 5, 2020).

Sandman, P. 2002. (February 21) Laundry List of 50 Outrage Reducers. https://www.psandman.com/col/laundry.htm (accessed June 5, 2020).

Sandman, P. 2007. (November 9.) Watch Out! How to Warn Apathetic People. https://www.psandman.com/col/watchout.htm (accessed June 5, 2020).

Technical University of Denmark. 2019. "Abundance of information narrows our collective attention span." ScienceDaily. (April 15.) www.sciencedaily.com/releases/2019/04/190415081959.htm (accessed June 9, 2020).

U. S. Bureau of Labor Statistics, https://www.bls.gov/iif/oshwc/cfoi/cftb0322.htm (data accessed February 8, 2020).

Weinstein, N., C. Cuite, J. Lyon, and P. Sandman. 1998. "Experimental Evidence for Stages of Health Behavior Change: The Precaution Adoption Process Model Applied to Home Radon Testing." *Health Psychology* 17 (5): 445–453.

Wojcik, S., and A. Hughes. 2019. Sizing Up Twitter Users. Pew Research Center. https://www.pewresearch.org/internet/2019/04/24/sizing-up-twitter-users/ (accessed June 28, 2020).

Postscript

This book was prepared to serve as both a grounding in sound principles and as a reference for navigating your way through risk communication efforts. It would not be complete, however, without a note regarding the unexpected.

Sometimes, despite doing all the right things, circumstances beyond our control still deprive us of our intended outcomes. Other times, impending disaster pivots on a twist of fate and the storm never hits. Oftentimes, though, we end up somewhere in between.

As Ryan Holiday writes in *The Obstacle is the Way*, under pressure and trial we get better. Those trials and pressures will inevitably come. And they won't ever stop coming. Regardless of the road you take, maintain your integrity, flexibility, and sense of humor. Even unexpected endings can bring us to great places.

References

Holiday, R. 2015. *The Obstacle is the Way: The Ancient Art of Turning Adversity to Advantage*. London, Profile Books Ltd.

*Appendix A: Laundry List of Outrage Reducers**

Giving Bad News

1. Tell people bad news they already know.
   If people already know something that is alarming to them and/or damaging to you—which is what I mean by "bad news"—you may feel there is not much point in telling them again. You are wrong. Confirming that yes, your factory belched some more smelly dimethyl meatloaf last night nails down the facts and squelches rumors. It also gets your credit for candor. If you can add useful detail, that is fine; if you cannot, just saying that is all you know so far is enormously better than saying nothing. The client I initially generated this list for is a railroad commuter. Passenger outrage when a train stops moving, this client has discovered, is greatly reduced when a staff person comes on the squawk box, even if all the staffer can say is that s/he does not know yet why the light is red or when it will turn green.
2. Repeat high-salience bad news again and again.
   People want to hear from you more than once about really significant mistakes or misbehaviors. In fact, they want to hear about them again and again—until they are sick of the topic; not until you are sick of it (which is instantaneous). Wallowing in damaging information is by far the best way to get past of it. We tire of it faster if you keep raising it than if we have to do the dirty work. (This is one of many examples of "the seesaw of risk communication.") Assuming Enron survives the current scandal, how long will it be before the company should issue a financial report without mentioning the bankruptcy of 2001?
3. Be sorry about bad news.
   Being sorry has three components. First, you must show that you regret it happened. (But do not use the word "regret"—it has become a legalistic weasel word.) Second, you must show that you sympathize with those who were damaged. (Again, "sympathy" is no longer a very sympathetic word.) Third and most important, you must apologize for your role in its happening. No credit for apologizing in the conditional. "I'm sorry if anyone was upset." "What do you

* Sandman, P. 2002. Laundry List of 50 Outrage Reducers. Reprinted with Permission from https://www.psandman.com/col/laundry.htm.

mean, if?" Apologizing requires knowing what you did and why it upset us. But apologizing does not have to mean acknowledging liability. It does not even have to mean acknowledging you made a mistake. Maybe it was not your fault and anyone else in your position would have fallen into the same hole. (At the start of the U.S. anthrax attacks, for example, the Centers for Disease Control did not realize that anthrax spores might escape a sealed envelope.) You have still got to be sorry.

4. Tell people bad news they are going to find out anyway.
   This is too obvious to belabor: Better we should hear it from you. But I am amazed how often my clients postpone and thus worsen the inevitable by waiting until an activist, journalist, or regulator breaks the bad news.
5. Consider telling people bad news they probably won't find out anyway.
   The toughest case for candor is when you can probably get away with secrecy. There are two reasons to think about telling the truth even in this situation (apart from law—I am assuming your lawyer says you do not have to tell). First, you might get caught after all. Damaging information generates roughly twenty times more outrage when you keep it secret and somebody else blows the whistle than when you blow the whistle on yourself. It follows that secrecy has to have a better than 95% success rate to cost out; if secrecy works only 80 or 90 percent of the time, candor pays. The second reason is subtler. When you reveal damaging information you did not have to reveal, you earn a reputation for transparency. People begin to notice that when you do something wrong, you say so; it follows that when you do not say so, you did not do anything wrong. "That odor can't be from the chemical plant or they would have called."
6. Tell people news that is not bad but might sound bad to them.
   The information my clients are least likely to give their stakeholders is benign information that does not sound benign. Your plant's solid waste includes a substance that is listed as a possible human carcinogen. But there is no exposure pathway, so there is no risk—none, zilch. You still need to tell people about it. If they find out any other way, they will be far less likely to accept your (belated) argument that the stuff is safe even though carcinogenic. Last year a Canadian chemical plant had a fire. It was readily visible to plant neighbors, but there were no significant toxic emissions, so rather than wake people up management simply put out the fire. The dialogue in the ensuing weeks and months was far more hostile than it would have been if management had used its alert system that night to tell neighbors the fire was under control.
7. If you have not given people much bad news in the past, explain the change.

Since most companies and government agencies serve up an unbalanced diet of mostly good news, the rest of us get used to adjusting. We assume you are only giving us, say, one-tenth of the bad news, so we multiply everything bad by ten. When you start being candid, it may take us a while to notice and stop multiplying. To short-circuit that awkward transition period, simply explain: "It's not that things are much worse than they used to be; it's just that we're telling you things we used to keep to ourselves."

8. Do not blindside anybody.
   Every time you have bad news to convey, think hard about whom to convey it to. Even when it is debatable whether to tell the general public, it should be a no-brainer to tell activists, political leaders, regulators, and other key stakeholders. They are the people who are likeliest to find out anyway, and likeliest to be angry that you did not tell them. Once you decide to tell some stakeholders, of course, you would better tell all the stakeholders who will otherwise be offended they were not on your A list – which means critics in particular. And if you are going to release the news publicly after all, give an early heads-up to your key stakeholders, and to anyone else the media are likely to call for comment (a local expert, for example). Tell your employees in advance as well, so they can explain things to their neighbors. Anyone you have blindsided will, when asked about what happened, choose to sound critical rather than sounding ignorant.
9. When explaining the causes of bad news, consider the stupidity defense.
   Whenever something bad happens on your watch, there are essentially four possible explanations for why it happened: (1) You were unlucky (the gods did it); (2) You were victimized (somebody else did it); (3) You were stupid (you did it by mistake); and (4) You were evil (you did it on purpose). Of course the truth—which is what you should be trying to tell—is usually a mixture, and either number three or number four is usually part of the mix. Usually it is number three. But unless you specify three, people are likely to jump to the fourth explanation. So if you really did not do it on purpose, admitting the ways you were stupid is crucial to keep us from imagining you were evil. "I can't believe we made such a boneheaded mistake!" You can use the stupidity defense even when bad luck and others' misbehavior were also part of the mix: "We should have found the problem quicker and been better prepared to deal with it." Lawyer won't let you say that? Try: "I wish we had found…."

Giving Good News

10. Tell people the good news too!
    I put so much emphasis on the importance of giving people damaging information, my clients sometimes imagine I think that is the only information they should give people. This misimpression does not seem to keep them from dishing out lots of self-serving information anyway. But it is a misimpression. Withholding reassuring information is a disservice to your stakeholders, as well as to yourself. I am not really worried that you might do it, but for the record: Do not withhold good news.
11. Acknowledge incredibility.
    Whenever you tell people something you know they are going to have trouble believing, tell them they are going to have trouble believing it. Tell them why—because it is self-serving; because common-sense says otherwise; because everybody else is on the other side; whatever. Acknowledge that the burden of proof is on you. Only after acknowledging incredibility can you productively try to meet that burden of proof and change their minds. It helps if you can honestly relate your own conversion: "That's what I thought when I first took this job. It took them weeks to convince me I was wrong."
12. Anticipate and acknowledge objections.
    In principle, good news for you should be good news for your stakeholders too ("our emissions won't give you cancer"). But embattled activists and outraged neighbors won't see it that way. They will have objections. Typically you not only know they will have objections; you know what their objections will be. You may be tempted to ignore their objections and just make your case. If you do that, they will be thinking more about their objections than about your case as they listen. Better to address the objections early and explicitly. Rebut the ones you can rebut (after acknowledging that it is natural to feel that way). Concede the validity of the objections that have some validity. Overall, talk less about your case and more about their objections and your responses.
13. Make the good news subordinate to the bad news.
    When people are worried or mistrustful (or both), good news is intrinsically not very credible. It becomes more credible when it is accompanied by the bad news, and subordinate to the bad news. "Even though this product has been approved by the regulators as safe for human consumption, there is a new study that suggests one of its ingredients could cause an allergic reaction in some individuals." "Even though the data say the building has been cleaned and workers can safely return, it's only human to worry whether there

might still be dangerous dust in the air." In the main clause you give people permission to be concerned; in the subordinate clause you explain why they probably do not have to be concerned. This is another instance of the seesaw. If you do it right, people should start telling you not to worry so much.

14. Acknowledge that self-serving information is self-serving.
    Whenever I tell clients that they should think about doing more risk communication training, I immediately add that they should take the recommendation with a grain of salt, since it comes from someone who sells risk communication training. One of the easiest ways to bolster the credibility of self-serving information is to acknowledge that it is self-serving.
15. Try to leave the good news for other sources.
    Third-party endorsements are another way of making the good news more credible; let it come from somebody else. Of course an endorser that is widely known as your ally or confederate won't help much (ask Arthur Andersen). A neutral third party is not bad, but the best endorsements come from your critics – which of course means that they will be half-hearted half-endorsements. A third-party quote in your corporate environmental report that says you are a wonderfully responsible company is worth only slightly more than saying so yourself. What you want is a quote that says, "I've been fighting with them for years, but I have to admit they're beginning to clean up their act." Your answer, ideally: "Well, thanks, but we've still got a long way to go."

Giving Uncertain News

16. Acknowledge uncertainty.
    There is no doubt that uncertainty exacerbates outrage; stakeholders would rather you were sure. But only if you really can be sure—that is, only if your certainty won't blow up in your face. Sounding certain and turning out wrong generates enormous outrage. Note also that expert disagreement leads to far more outrage than mere uncertainty; it follows that even if you are pretty sure you will turn out right in the end, you can afford to sound certain only if nobody credible is going to disagree. The rest of the time you must explain not only what you think, but also your understanding of what others think, and why the disagreement exists, and what you are doing to learn more. Above all, you must explain just how uncertain you are, taking (and justifying) a position somewhere along the dimension between total certainty and total ignorance.

17. Don't wait till you're certain.
If sounding certain and turning out wrong generates so much outrage, and even acknowledging uncertainty generates a fair amount, why not just wait till you are certain? That makes sense if interest, outrage, and hazard are all low; if nobody else is going to express an opinion while you are hesitating; and if it won't be long till you know for sure. The rest of the time, which is most of the time, you cannot afford to wait. In the wake of a potentially serious accident, for example, you cannot just tell people you will have a report out some time next spring. Like a doctor with an uncertain diagnosis, learn to explain the alternatives: "We think it's probably X, but it might turn out to be Y or Z, or even something else we haven't considered yet. Here's what we'll do about it if it is X. Here's what we'll do if it's Y or Z. Here's how we plan to find out more. And here's what we're doing in the meantime."
18. Do dilemma-sharing.
Dilemma-sharing is a special case of uncertainty—uncertainty about what to do. As with any other sort of uncertainty, the principle is clear: If you are not sure, show you are not sure. This is most obvious when you have not decided what to do. Share the dilemma and seek advice. But dilemma-sharing is even more valuable when you have decided what to do, but it was a close decision. If you do not want critics to claim your decision was obviously the wrong one, be sure not to claim it was obviously the right one. "We may be wrong" puts the debate on a much more reasonable, moderate path; it also saves your bacon if you turn out wrong in the end.
19. Show reasonable confidence when acknowledging uncertainty or sharing dilemmas.
Just because you are not sure does not mean you have to be quaking in your boots. Unless you actually are quaking in your boots, make sure you present your uncertainty confidently. (Don't use the word "confident." It sounds too certain. You want to sound uncertain, but not distraught.) I am not urging you to send a double-message, claiming uncertainty with your words and certainty with your tone. Rather, I am recommending that you model for your stakeholders an ability to tolerate uncertainty and still function.
20. Stake out the middle ground.
The dynamics of outrage are not symmetrical. Whenever you are trying to increase people's outrage, exaggeration is a useful tool. There are limits beyond which the claim lacks credibility (and ethics), but within those limits the more you exaggerate the more outrage you will get. When you are trying to diminish people's outrage, on the other hand, exaggeration backfires. It follows that most well-fought controversies are battles between the outrage-increasing extreme and

the outrage-reducing middle. If the fight is between their extreme and your extreme—that is, if they con you into a polarized debate—they win. Those of you who yearn for the pleasures of exaggeration, go join an activist group. When playing defense, exaggeration has no place in your playbook. Stake out the middle ground.

21. Don't claim anything is "safe."
    Even more than other sorts of controversies, risk controversies are not symmetrical. Just like professional risk assessors, citizens judge risk conservatively; they would rather be unnecessarily cautious than dangerously cavalier. For this reason (and some others), those on the alarming side of a risk controversy can afford to make extreme statements, but those on the reassuring side cannot. In a fight between "perfectly safe" and "incredibly dangerous," "incredibly dangerous" is a sure winner. But in a fight between "a little dangerous" and "incredibly dangerous," "a little dangerous" is a contender. So don't claim anything is "safe." It is safer than something else, or safer than some standard, or safer than it used to be. But it isn't flat-out "safe."
22. Tolerate exaggeration from opponents.
    People on the alarming side of a risk controversy, or the outrage-arousing side of any controversy, are on strategically solid ground when they exaggerate. People on the reassuring, outrage-reducing side—your side—are wise to stake out the middle. So when you accuse critics of exaggerating how big the problem is, you come across as defensive and dishonest. Even if you are right and we know it, we see their exaggeration as protective; we do not resent it as much as we resent your pointing it out. (But if they catch you exaggerating how small the problem is, they have got you nailed. The controversy is not symmetrical. Get used to it.) When a passenger complained to my commuter railroad client that the train was 27 minutes late yesterday, the customer relations office responded with proof that the train was only 23 minutes late. Wrong move. Here is a far better answer: "I bet it felt like an hour and 27 minutes!"
23. Ride the seesaw. In the long run, aim for the fulcrum.
    I have already mentioned the risk communication seesaw, but it deserves its own place on the list. Whenever people are ambivalent—which is not always, but it is more frequent than you might think—they tend to emphasize whichever side of the ambivalence you are neglecting. If you say the risk probability is low, they will say the magnitude is high. But if you stress how many people you might kill if things go really wrong, they will stress how vanishingly unlikely that is. And if you move to the middle of the seesaw, balancing two conflicting realities and enduring the ambivalence, so will they. There are lots of seesaws in outrage management, among them blame, preparedness, cost-benefit, and certainty as well

as risk. First determine if your stakeholders are ambivalent. If they are, decide where on the seesaw you want them, and locate yourself in the counter-balancing position. Then work your (and their) way toward the fulcrum.

Listening

24. Talk less; listen more.
Well, we have covered 23 recommendations already. The column is getting long, I am getting tired—and so are you—and so far it sounds like I think outrage management is all about talking better. My mistake. In managing people's outrage, what you say is less important than what they say. Shut up and listen.
25. Let people blow off steam.
When outrage is high, people's foremost need is to vent. Even apologizing won't do any good until they are done telling you how angry they are. (Married couples know this.) In really high-outrage situations, therefore, focus on creating opportunities for venting. If you face scores of outraged people, set up and endure a hostile public meeting. Do not try to make any points; in fact, the first few times anyone challenges you for your answer, demur on the grounds that it seems to you people have more to say first. (Yes, this is the seesaw again.) If it is just one or two people who need to vent, get them an invitation to your Community Advisory Panel. And if necessary quarrel (gently) with the panel in defense of their right to express their views … again and yet again. (Yep, the seesaw.) Unless there is a safety emergency, the expression of stakeholder outrage should always preempt anything you might have to say.
26. Listen actively.
When you listen, listen actively. An occasional uh-huh is better than nothing, but you can do better than that. Take notes (but not constantly; take time to face your accuser as well). Periodically try to summarize what you think you are hearing, and ask if you have got it right. (For major meetings the notes should be on a flipchart and the summary should be written up and distributed the week after.) Look thoughtful, sympathetic, or penitent as appropriate; do not look bored or frozen. Ask questions—clarifying questions, not hostile questions. Of course it is possible to do all these things in a way that strikes the rest of us as sarcastic or threatening. Your goal is not to go through the motions of listening while managing to convey

that you do not think much of what you are hearing. Your goal is to get it … and show that you are getting it, or at least that you are trying hard to get it.

27. Try to connect to what you are hearing.
    In the literature this is sometimes called "empathic identification." It goes beyond active listening to find a way to connect to what you are hearing. Remembering and using people's names is a simple example. Retelling their stories is another. (An answer to a complaint letter should always summarize the complaint, preferably in detail. "You feel we did three things wrong….") A more complicated version is trying to describe the issue at hand as your stakeholders see it, rather than just as you see it. For you, that leaking hazardous waste site may be about engineering: dikes and lysimeters and the like. For them, it is about sinking property values and that sinking feeling in the pit of the stomach when someone says their children may get cancer in 20 years. At the most basic level, people would rather you told them about themselves—their experiences, their problems, their fears—than about you and your company. But empathic identification can backfire; read the next paragraph.
28. Be careful how you connect to what you are hearing.
    My clients sometimes try to connect to what they are hearing in ways that backfire because they sound too much like the teacher praising a student for a good answer. "You're right" may be a better response than "you're wrong," but it is not as good as "I see what you mean." Even "I see what you mean" can claim too much – "I know how you feel" is almost guaranteed to yield a "No you don't!" Paradoxically, people feel more understood when you empathically say you do not understand than when you smugly say you do; try murmuring "I can't really imagine what that must be like." For similar reasons, think twice before you tell a story about how you once faced a similar situation. This might be the perfect way to show you get it—or it might feel like you are changing the subject from their grievances to your life history. Do not steal the stage, and do not claim too much understanding.
29. Find indirect ways to raise what you are not hearing.
    Often you get a sense that there is something on people's minds they are reluctant to mention—and until it is addressed somehow, you are not going to be able to make things better. One very common example in risk controversies is people complaining about health risk who are actually more worried about economic risk. Even more common, in all sorts of controversies, is people whose apparent outrage at you is rooted in their own injured self-esteem. Whatever the unacknowledged issue, almost by definition you cannot raise it directly without giving offense. So you need to find

indirect ways to get it into the room, if not quite onto the table: "Some people might be worried about…." "I talked to someone last week who felt…." "I think if I were in your shoes the thing that would get to me is…."

30. Be human.
    As community people get more and more passionate about a risk controversy, government and corporate experts tend to get more and more dispassionate. This forces both sides into caricatures of themselves: the uncaring technocrat versus the hysterical neighbor. Instead of dispassion, aim for compassion. And if compassion is not attainable, find in yourself a human response that is attainable, and go with that. When you are accused of spreading disease through the neighborhood, even angry denial is better than calm absence of concern. More generally, look for ways to humanize your interactions. Eat together. Talk about other things during breaks. Ask after their families, and tell them about yours. You do not want to be unprofessional, and you certainly do not want to trivialize your critics' concerns. But you do want to be a person.
31. Have a substantive response.
    I have focused on the non-substantive side of listening. But substance obviously matters too. Ultimately, the most important way you show you have been listening is by changing in response to what you heard. Genuinely useful things are said in virtually every public interaction. They may be things you never thought of; more often they are things you considered and put aside—because there were higher priorities at the time, because someone inside your organization was opposed, whatever. The trick now is to put aside your own ego (and your management's ego). Instead of looking for ways to tell people their suggestion won't work, or claiming you were going to do it anyway, look for ways to adopt the suggestion or some version of it—and say so. Of course even explaining why a suggestion won't work is more respectful and responsive than pretending it never came up.
32. Take warnings seriously.
    The public comments that most require a substantive response are warnings that something bad might happen and you are insufficiently prepared to prevent it or cope with it. Every warning deserves an answer. And every warning that raises new questions deserves a reconsideration of the old answers. Obviously that does not mean that you can afford to take every precaution your stakeholders can think of. But you can afford to think the issue through again and again, with your stakeholders' help, trying to figure out if there are feasible and cost-effective approaches that will reduce the risk. Here is a good way to make sure you do not under-respond to warnings:

Imagine that the possible outcome you are being warned about has actually happened, and you must explain to the media and the regulators how you dealt with the warning.

Collaborating

33. Ask for help.
    Benjamin Franklin once wrote that the best way to make a new friend is to borrow his pen. Franklin was right. Lean on your stakeholders as much as you can—and the more critical they are, the more you should lean on them. Identify problems you have not solved yet and ask for your stakeholders' advice. Describe choices you face and ask them which option they prefer … or whether they can think of another. Ask them what they remember about where the barrels were buried back in the fifties; ask them to keep a log of strange odors and where they seem to be coming from. Of course the tasks you set your stakeholders should be real tasks – do not get them working on a project you have no intention of using. But do not worry about asking for too much. A neighbor who wants you to back off and run your own factory is a big improvement over a neighbor who demands the right to run it for you. (There is that seesaw again.)
34. Learn things together.
    Every parent and primary school teacher knows that experiential learning is the best kind; we learn better by doing than by watching. This is much, much truer when the material to be learned is controversial, the "teacher" has a stake in the outcome, and the "students" are already mistrustful and upset. If you want your toxicology or epidemiology study to be credible, therefore, do not do it yourself (or hire it out) and then report the results. Do it collaboratively with your stakeholders. Set up the research so you could not cheat even if you wanted to, and you will make it very hard for the rest of us to claim you cheated. (Of course you will also make it hard for you to cheat if you do not like the results; that is the point.) The same thing is true on a moment-by-moment basis. Do not rush to provide a piece of information if you can wait till we get it off the Internet instead. Better yet, get it off the Internet with us … and make sure it is a stakeholder, not you, running the search.
35. Share control.
    People are much less outraged by risks they control than by risks you control. That is why nearly everyone feels safer behind the wheel of a car than in the passenger seat. It follows that to reduce people's

outrage you should share the control. Of course there are some decisions you are unable or unwilling to share; it is your company. (Even in these cases you can ask for input and take it seriously.) On the other extreme, there should be decisions that matter much more to your stakeholders than they do to you, where you can go beyond sharing control to delegating control. In choosing which decisions fall where on this dimension, try to distinguish the promptings of your ego from the necessities of your business. And do not forget to weigh the business value of reducing stakeholder outrage.

36. Be accountable.
    Sharing control reliably reduces stakeholder outrage, but unfortunately it reliably increases corporate outrage. Accountability is a good compromise. If sharing control requires that you let stakeholders actually help drive the car, accountability requires only that you let them be backseat drivers, watching and criticizing as you drive. Accountability is also the best answer to mistrust. Instead of trying to get us to trust you, a difficult and dubious goal, try instead to make what you do more accountable, so we do not have to trust you; we can check for ourselves. Once you install satellite emissions monitors in the lobby of the community rec center, neighbors will stop asking whether you are telling the truth about your emissions.
37. Give credit where credit is due.
    My clients are usually better at sharing control and being accountable than they are at giving away the credit. They actually do respond to stakeholder concerns, then paradoxically insist that "we were going to do that anyway." This is of course irritating to the people who deserve the credit; it lessens the likelihood that they will move on to another interest and increases the likelihood that they will attack your response as too little too late. It also saps the credibility of the response itself. If you say you are doing the right thing because you want to, our skepticism will be high; if you say you are doing the right thing because we made you, we are far likelier to think you are doing the right thing. Like many outrage management strategies, giving away the credit is bad for your ego but good for your business.

Other Important Outrage Reducers

38. Do anticipatory guidance.
    Anticipatory guidance is information on what to expect. Anyone who has ever followed a set of directions knows how reassuring it

is to read, "Then you'll see a Mobil station on your left" … and sure enough there's the station. "It will take us about three weeks to look into the issues you raised in your letter" is anticipatory guidance on when to expect an answer. "The results of the epi study probably won't give us a definite answer, and if you're like most people that could be disappointing, even infuriating" is anticipatory guidance on what is likely to happen and how it is likely to feel. Anticipatory guidance can be good news, bad news, or uncertain news. What matters is that it's early news. Knowing what to expect makes the wait easier to endure. It gives people time to get ready for what is going to happen. When it happens, we can deal with it better. Whether the news is good, bad, uncertain, or indifferent, we all benefit from knowing in advance. You benefit from telling us.

39. Don't be overoptimistic, and don't overpromise.
Anticipatory guidance is best, obviously, when you turn out right. But some ways of being wrong are much worse than others. If you say things are going to be fine, and they turn out bad (even modestly bad), outrage is magnified and credibility is destroyed. If your predictions err on the pessimistic side, on the other hand, people are relieved, even though they may blame you for scaring them unnecessarily. So err on the side of caution. Make sure things never turn out worse than you said, even at the cost of having them regularly turn out not as bad as you said. This is not just true of risk estimates; it is true about all anticipatory guidance. Better to predict a site remedy in three years and have it in two than to predict it in two and have it in three. One especially important implication concerns predictions about your own behavior. It is tempting when apologizing to promise "We'll never do it again." Fine – if that is a promise you can keep. Otherwise, aim lower.

40. Use counter-projection.
People's responses to an issue can be distorted by things they are not letting themselves say, not even letting themselves think. If you sense one of these unacknowledged distorters getting in the way, try to reduce its impact by bringing it closer to the surface—but indirectly, so you won't force it deeper instead. That is one kind of counter-projection. In the recent anthrax attacks in the United States, millions of people felt fear heightened by hurt ("someone out there wants to kill me"), misery ("in another attack I would again be forced to watch horrific things on TV news"), empathy ("we're all in this together"), and rage ("revenge!"). Alleviating the fear required addressing these other emotions that were masquerading as fear … but without accusing people of harboring them. "Some people tell me they feel…." A simpler example: People who mistrust you just because you work for a large corporation will not begin to listen until

you back off the substantive issue and address the awful reputation of large corporations.

41. Express hopes and wishes.
I am amazed at how seldom my clients tell their stakeholders their hopes and wishes. This is of course humanizing—it establishes that you are a person who has hopes and wishes. If they are hopes and wishes your stakeholders share, then expressing them builds commonality. And if they are hopes and wishes your stakeholders share but cannot acknowledge, expressing them becomes a tool of counter-projection. Achievable hopes are worth expressing; in fact, every time you are tempted to make a promise, consider whether you would be wiser to downgrade it to a hope. But unachievable wishes are even more valuable to express: "I wish I could give you a more definite answer." "I wish that damn accident had never happened." Think about how much better these statements are than "I simply cannot give you a definite answer" or "I know you wish the accident had never happened."
42. Be careful with risk comparisons.
Outrage and hazard are independent components of risk. When you compare a high-outrage low-hazard risk with a high-hazard low-outrage risk, your point is usually that the first is less risky than the second. To choose a classic example: "Our emissions are a smaller risk than the risk you took driving to this meeting." In hazard terms, you may well be right. But people will hear you in outrage terms, and in outrage terms it just ain't so. Your emissions are much higher-outrage than driving. As a rule of thumb, simply do not compare the hazard of two risks whose outrage varies in the opposite direction. Or if you do, bend over backwards to acknowledge that you understand why your emissions are more objectionable than driving even if they are not more dangerous, that you realize the acceptability of a risk is a product of more than just its hazard, that you are not trying to corner people into giving your emissions a free pass just because they are unwilling to give up their cars.
43. Acknowledge acknowledge acknowledge.
I have used the word "acknowledge" a lot in this list already. Like "location" in real estate, acknowledgment is the core of managing outrage. Acknowledge what went wrong; acknowledge what your stakeholders think went wrong; acknowledge what it is about you that makes them want to think so. Acknowledge everything that disposes your stakeholders to be against you. In many other kinds of communication—consumer marketing, for example, or political campaigning—it is wise to focus on selling your strengths. Even public relations assumes the audience is not listening closely and critically enough to justify much acknowledgment. True, when the

audience is on your side, or at least neutral, or when it is not very interested and not likely to learn much, a little acknowledgment goes a long way. But when the audience is already against you, or is likely to listen hard to your opponents later, then you are doing outrage management—and acknowledgment is the name of the game.

Everything Else**

44. Monitor and manage your own outrage.
    My clients like to imagine that risk controversies are battles between outraged communities and calm companies. But when people are questioning your competence and your integrity, accusing you of killing their children, and costing you millions of dollars to boot, you are bound to be outraged too. In fact, the single biggest barrier to managing other people's outrage is your own outrage. At its most basic, outrage management is a set of strategies for profitably losing fights—fights you may think you ought to win, fights you may want to win more than you want to make money. The first step in managing your own outrage, of course, is monitoring it. Be particularly alert for coldly calm, courteously unhelpful, deniably hostile patterns of behavior, where everybody except you knows you are angry and you imagine you are just right. Psychiatrists call this pattern "passive-aggressive." It is a sure sign of unacknowledged outrage.
45. Distinguish outrage from greed.
    Bargaining works best when both sides are greedy and neither is outraged; both sides want to win, so a win-win is the optimal outcome. As soon as anyone is outraged, winning is no longer their goal; they want you to lose. Effective strategies for coping with greed backfire when you are facing outrage instead. For example, a cash offer that feels like the start of negotiations to a greedy opponent will feel like a bribe to an outraged one—and will therefore worsen the outrage. But the reverse is also true: Outrage management won't work well when greed is the main thing going on. Apologies and concessions that can calm an outraged opponent, for example, may strike a greedy opponent as signs of weakness. In diagnosing what mix of outrage and greed you face, remember that it is far more common to imagine your stakeholders are greedy when they are actually outraged than to imagine they are outraged when they are actually greedy.

** Besides Stakeholder Outrage

46. Protect stakeholder self-esteem.
    Just as outrage often looks like greed, injured self-esteem often looks like outrage. Homeowners near a controversial factory may feel guilty that they moved their families into what now looks like a dangerous situation; they will convert their guilt into outrage at the company. Many recipients of breast implants felt ashamed of their breasts, their implants, their mastectomies; they converted their shame into outrage at the company. When a young child had a seizure on a client's bus, the flustered mother asked the driver to let them off so she could call for medical help. Of course he should have noticed the child's condition and taken action himself. But the mother's later outrage at the bus company was fueled by her embarrassment at having mishandled the emergency. In all these cases, you cannot reduce the outrage unless you first reduce the ego damage. Better yet, try not to cause ego damage. People who feel bad about themselves are likely to take these feelings out on you.
47. Watch your own self-esteem.
    Managing other people's outrage (and their greed and self-esteem) may save your company a lot of money, but it probably won't go easy on your ego. Saying you are sorry, giving away credit, acknowledging mistakes and misbehaviors, not overpromising—nearly all the recommendations in this column, in fact, can be threatening to the self-esteem of the manager who uses them. And these recommendations tend to be most needed just when your self-esteem is at its most vulnerable: when your company has messed up, when your stakeholders are hurling hurtful accusations at you, when your own management is wondering if you know what you are doing. Inside an organization, good managers learn not to lash out at subordinates, colleagues, or supervisors just to make themselves feel better. The same self-knowledge and self-restraint will help you deal with external stakeholders.
48. Look for ideology and revenge.
    Outrage, greed, and self-esteem are the "big three" in risk controversies (and most non-risk controversies too). But sometimes stakeholders are motivated instead, or as well, by ideology or revenge. Regulators and full-time activists are the likeliest to have ideological objections to your company's behavior (or its very existence). When you encounter ideological opposition, it may or may not be useful to point it out—to the opponent, the opponent's management, or the public – but it will certainly be useful to know it is there. As for revenge, I am struck by my clients' tendency to discount this normal human motivator. You have every right to go over a local regulator's head and get the politicians to take the pressure off. You have every

right to make a critical neighbor look stupid in the media. But then do not be surprised when they get even.

49. Look for fear.
In most of the controversies I work on, what looks like fear is actually mostly outrage. If you take action to reduce the outrage, you can expect the fear to diminish on its own. Except when fueled by outrage, in fact, excessive fear about public issues is unusual; even appropriate fear is unusual. When outrage is low, the normal problem is apathy. But there are exceptions, times when people are more afraid than angry or outraged. Even then, over-reassurance is likely to backfire. When people are frightened, the job is to help them bear their fear, not to talk them out of it. That means acknowledging that the fear is natural, and if possible it means suggesting things they can do to lessen their risk. What happens when people simply cannot bear their fear? On rare occasions, panic – but only on rare occasions. The more usual response to unbearable fear is denial. Watch for denial, and try not to mistake it for apathy. Apathetic people need to be shook up; those in denial need a gentler response.

50. Don't forget hazard!
I partition risk into two components—the non-technical component I call "outrage" and the technical component (what experts mean by risk) I call "hazard." This column has focused on ways to manage outrage. That does not mean you should ignore the hazard. When outrage is high and hazard is low—which is common—there is not much to be done about the hazard; managing the outrage is your main task. In this high-outrage low-hazard situation, over-managing that tiny hazard is an expensive, inefficient way to reduce the outrage; you will be better off with the recommendations in this column. In a high-hazard low-outrage situation, obviously, all you need to do is manage the hazard. But often both hazard and outrage are high enough to deserve serious attention. It is irresponsible (and ultimately ineffective) to manage the outrage instead of the hazard in such situations. Manage both.

Appendix B: Environmental Risk Communication Archives—Fallout from Three Mile Island (TMI)

It was the story that just would not go away. Or so it seemed. First there was the accident on March 28, 1979. Then came the clean-up stories, the undamaged Unit One re-start hearings stories, the anti-nuke protests, and one-year anniversary stories, the first truck of radioactive waste leaving the plant stories, more clean up stories, more anti-nuke protests, more company updates, and so on.

Three Mile Island (TMI) incident is called the nation's worst nuclear accident. To media observers and public relations practitioners it was a communications disaster. One of the goals of a good crisis communications plan is to contain the duration of the accident and help get operations back to normal as quickly as possible. The TMI accident redefined normal and as a result, the entire commercial U.S. nuclear power program was redefined and the way corporations and government approached crisis communications changed.

Following the TMI accident, the power "too cheap to meter" went under intense scrutiny. The way company officials handled the flow of information was debated and in Washington, DC, investigations were held to make sure a repeat of TMI did not occur. Things would never be normal at TMI. One of its two reactors was too damaged to ever go back on-line and in 1979, the future of the undamaged Unit One reactor was unclear.

Prior to the accident, TMI was this place that most people in Harrisburg did not care much about. Except maybe for the construction workers lucky enough to get jobs there. As far as news from the Island, there was not much. A scan of the *Harrisburg Patriot News* published in the months preceding the plant opening had little to say about TMI, nuclear power, emergency evacuation plans, potassium iodine, or anything else—until March 28, 1979.

The series of events surrounding the accident have been well documented. During the early hours of the accident, TMI Unit Two plant operators worked to figure out what the instrument readings meant. Inside the reactor, temperatures rose and the start of the worst U.S. nuclear power plant accident was underway.

On that late March morning, commuters travelled from their suburban homes to the Capital Complex in downtown Harrisburg or one of the two large military depots that provided good jobs for the residents of Central Pennsylvania. As workers shuffled into work, school buses were unloading hundreds of kids. Restaurants were filled with morning regulars, stores were unlocking their doors and movie theatre marquees advertised "The China Syndrome," starring Jane Fonda and Michael Douglas.

Things were pretty much normal that morning, until "Captain Dave Edwards" a radio traffic reporter for Harrisburg radio station WKBO noticed that there was not any steam coming from the huge TMI cooling towers. Unit One was already shut down, so the remaining two cooling towers would normally emit a tower of steam visible for miles around. Captain Dave also received a citizen band radio call from a friend who told him fire companies were being put on alert. Captain Dave picked up the radio and called his newsroom with the information [Stephens, 1980].

The very minute Captain Dave keyed his radio microphone to report what he had learned marked the beginning of a new era in journalism and corporate crisis communications. Back in the newsroom of WKBO radio, news director Mike Pinteck phoned TMI and was transferred to the control room by the operator. "We can't talk right now, we've got a problem," responded the voice on the other end of the line [Stephens, 1980].

The employee whose job description most closely resembled that of a public relations professional on duty at the Island on March 28, 1979 was a tour guide at the visitors' center. The utility's primary public relations staff was located in Reading, PA approximately 60 miles from the island. Within hours of WKBO's report, an estimated 400 reporters, photographers, and producers descended on Middletown, PA. Hundreds of reporters vs. one tour guide.

The crisis communications lessons learned from the TMI accident are considered "textbook" examples of what not to do during a crisis. The utility did not have a system in place at the island to deal effectively with the media, and few reporters understood nuclear power.

There was a lack of crisis communications planning. News briefings were held by state officials at the Capitol building in Harrisburg and eventually the utility utilized an American Legion social hall to hold its briefings. The release of public information was not coordinated between the utility and the state and at times, information was contradicted by one source or the other, causing complete confusion for the general public. Eventually the state would not consider the utility information credible and would only release information that it gathered from its own sources [Stephens, 1980].

President Jimmy Carter dispatched Harold Denton from the Nuclear Regulatory Commission to the scene. Mr. Denton, at the urging of his public affairs staff, did not issue joint news releases with the utility.

Lessons Learned from TMI

Following the TMI accident there were several federal investigations by the Nuclear Regulatory Commission, Environmental Protection Agency, and the President's Commission on the Accident of Three Mile Island referred to as the Kemeny Commission. Recommendations from those investigations lead to the formation of regulations on emergency planning for nuclear power plants, including mandates for alert and notifications, public warnings, and public information [Golding et al., 1995].

Beyond the mandates of the federal regulations, the public relations industry studied the lessons learned from TMI and redefined its own policies for crisis communications. Today, the Commonwealth of Pennsylvania Emergency Management Agency (PEMA) has in place a state-of-the-art emergency operations facility that would serve as the center for information and decision making in the event of another nuclear incident or any other type of major disaster.

As a result of the Kemeny Commission, the Federal Emergency Management Agency (FEMA) now has the responsibility for off-site emergency planning for nuclear power plants. Nuclear utilities are now required to test their emergency plans, including emergency communications procedures biennially and conduct an on-site drill annually.

By the end of the TMI incident, a near-site media center was set up, which was complete with communications equipment to brief the throng of reporters who might once again descend on Middletown. The center was a far cry from the crowded American Legion social hall.

TMI has taught many in the public relations industry the importance of a good, well-tested, crisis communications plan. Company officials admit that

Reporters at TMI. Immediately after the Three Mile Island story broke, an estimated 400 reporters descended on Middletown, Pennsylvania. The utility was unprepared for such a large group of reporters. TMI set up a media center to facilitate media briefings.
(Photo reprinted with permission from The Historical Society of Dauphin County, The John Harris/Simon Cameron Mansion Museum, Harrisburg, PA.)

the spokespeople used during the accident were not trained in media relations and their performance and ultimately the company's reputation suffered.

References

Golding, D., Kasperson, J. X., and Kasperson, R. E., Eds. 1995. *Preparing for Nuclear Power Plant Accidents,* Boulder: Westview Press.

Stephens, M. 1980. *Three Mile Island: The 'Hour-by-Hour Account of What Really Happened.* Rev. ed. New York: Random House.

*Appendix C: Gauging Level of Public Interest in a RCRA Facility**

Level of Public Interest	Type of RCRA Action	Community's Relationship With Facility/Agency	Larger Context
Low	• The RCRA activity is unlikely to be controversial (e.g., a routine permit modification). • There is no contamination at the facility that could come into direct contact with the public.	• People do not live near the facility. • There is a history of good relations between the facility and community members. • Community members have expressed confidence in the agency and/or facility.	• The facility receives very little media attention and is not a political issue. • Community members have not shown any past interest in hazardous waste issues. • Public meetings on the permit modifications have not been attended.
Moderate	• The RCRA action may involve activities that contribute to a public perception that the facility is not operating safely. • Examples may include permits for storage and on-site activities or routine corrective actions. • Highly toxic and/or carcinogenic waste may be involved (e.g., dioxins).	• A large number of people live near the facility. • There is a history of mediocre relations between the facility and community members. • The facility is important to the community economically, and the action may affect facility operations. • Community members have had little or poor contact with the agency. • Local elected officials have expressed concern about the facility.	• Community members have shown concern about hazardous waste issues in the past. • The facility receives some media attention, and there are organized environmental groups interested in the action. • There are other RCRA facilities or CERCLA sites in the area that have raised interest or concern.

(*Continued*)

* **U.S. EPA Resource Conservation and Recovery Act Public Participation Manual, 2016, 530-R-16-013.** https://www.epa.gov/sites/production/files/2019-09/documents/final_rcra_ppm_updated.pdf

Level of Public Interest	Type of RCRA Action	Community's Relationship With Facility/Agency	Larger Context
High	• The RCRA action includes a controversial technology or is high profile for other reasons (e.g., media attention). • Highly toxic and/or highly carcinogenic waste is involved (e.g., dioxins). • There is potential for release of hazardous substances or constituents that pose potential harm to the community and the environment. • There is direct or potential community contact with contamination from the facility (e.g., contaminated drinking water wells or recreational lake).	• The nearest community population is within a one-mile radius. • A large number of people live near the facility. • There is a history of poor relations between the facility and the community. • The facility has violated regulations and community members have little confidence in the agency to prevent future violations. • There is organized community opposition to the facility's hazardous waste management practices or to the action. • Outside groups, such as national environmental organizations, or state or federal elected officials have expressed concern about the facility or action. • The economy of the area is tied to the facility's operations.	• Community members have shown concern about hazardous waste issues in the past. • Facility activities are covered widely in the media. • There is interest in the facility as a political issue, at the local, state, or federal level (e.g., statewide or national environmental groups are interested in the regulatory action). • There are other issues of importance to community members that could affect the RCRA action (e.g., concern over a cancer cluster near an area where a facility is applying for a permit to operate an incinerator). • There are other RCRA facilities or CERCLA sites nearby that have been controversial.

*Appendix D: Examples of Bridging Statements**

Personal Opinions

Q: "What do you know about X?"

A: "I don't know about X, but what I can tell you about Y is..."

Q: "Critics say X about your organization."

A: "I can't speak for them, but I do know that..."

Q: "Agency X has said...Do you agree?"

A: "I can't speak for X. What I can tell you is... (If you are not responsible for what the media is referring to.)"

A: "I agree that... (If you are responsible for what the media is referring to.)"

Guarantee/100% Assurance

Q: "Can you guarantee this will never happen again?"

A: "What I can guarantee is...Let me give you one (or two) examples." A: "We have conducted extensive..."

A: "We will do everything it takes to investigate..."

The Set Up

Q: "If it is determined that...will you pay for...?"

A: "I can't speculate. Now, what I can tell you is...," "Our policy requires..."

Q: "What if your employees are found negligent?"

A: "A thorough review is underway right now. I can't speculate on any causes right now."

False Choice

Q: "Isn't it better to be safe than sorry?"

A: "We take safety very seriously. For example..."

---

* U.S. Navy. Undated. Navy and Marine Corps Public Health Center. *Risk Communication, Appendix L. Examples of Bridging Statements* Fulton, K., and S. Martinez, Fulton Communications.

Hypothetical/Rumor/Speculation

Q: "What if…"

A: "I can't speculate, but I can tell you that…"

A: "That's a hypothetical question, but what we do know is…"

A: "Unfortunately, we don't have a crystal ball. What I can say is…"

Q: "Isn't it possible that this could have been caused by…"

A: "Here's what we know right now…"

Q: "We've heard that there's a possibility that this was caused by…"

A: "What we know at the moment is…"

Multiple Choice (Pick Your Position)

Q: "So, what is your track record and have you in fact dealt with this problem before and do you accept responsibility for this?"

A: "Well, to answer your first (or second or third) question…" (Only respond to the questions that apply to your messages and pick the one you want to answer).

Foot-in-Mouth

Q: "So you would say (your organization) has a far better reputation than…"

A: "What I AM saying is…"

A: "What we are proud about is…"

A: "Our organization has accomplished…"

False Premise/Negative Allegation

Q: "There's clearly been a cover-up. Isn't it time you came clean with the local community?"

A: "We've been very open about our operations. In fact, we have…"

Q: "Didn't your organization lie about…?"

A: "We've been truthful about…. For example…"

Q: "What if it is determined that…," "You intentionally…"

A: "What we are doing is…"

Speculate Based on Past Events

Q: "Didn't this happen before and was it due to operator error?"

A: "At this point, we don't know what caused this incident. That's why we are conducting an investigation…"

A: "We learned a lot from the past incident and made improvements."

How to Defer to the Correct Person

A: "I know you want updated information. The person who can give that to you is…"

A: "I don't have the information you want. 'X' can give that to you. Here is the phone number."

A: "I'm 'x' and my job is to 'y.' The person who can help you is…"

Appendix E: Laundry List of Media Interview Tips

- Nothing is ever off the record, including the polite banter before and after the "official" interview.
- The sooner you return a reporter's call, the better chance you have of getting your message covered and helping to shape the coverage.
- *Never* repeat the negative language in a question. Instead begin with a statement such as "That's incorrect..." or "I don't agree with your premise."
- Never say "no comment."
- Avoid humor. It fails many more times than succeeds.
- Be critical of the issues, not critics or competitors.
- Do not use absolutes. Chances are they will come back to bite you.
- Always tell the truth.
- Prepare, prepare, prepare.
- Practice, practice, practice.
- Remember that it is *not* your job to provide balance. That is the reporter's job, and the reporter is likely to achieve balance by speaking with someone having an opposing view.
- If you are inexperienced, seek training. Record yourself on video to help achieve objective feedback.
- Learn when to STOP talking. Continuing to speak after answering the question risks going off message and creating gaffes.
- Record the interview when you perceive a high risk of being misquoted or misrepresented (for example, a hostile or agenda-driven interviewer). (If you do plan to record the interview, let the reporter know in advance.)
- Mind the visuals for video. This includes your background, your makeup, your clothes, your body language, potential sources of distraction, etc.
- Be professional and keep your cool—always.
- Smile. Look interested and alert.
- Speak more slowly and more loudly. Nervous people tend to speak quickly, and news spokespeople often speak too softly.

- Do not be afraid to ask for clarification or to say "I don't know the answer to that, but I will find it out and get back to you."
- Speak in a natural and plain manner. Do not try to impress the audience with your knowledge.
- Do not try to memorize all possible responses to questions. Remember the most important information to be conveyed and practice thinking on your feet in advance.
- Maintain eye contact with whomever is asking the questions.
- Do not rush to fill silence. Reporters may use silence to encourage you to keep talking and stray off message. Keep eye contact. If silence keeps dragging on, you could ask if he or she has any additional questions.
- Never comment on what others have said, particularly if you have not seen or heard it. Do not verify something that might not be true.

Appendix F: Laundry List of Tips for Giving an Effective Presentation

- Define your goal.
- Know your three key points.
- Know your opening and closing. You will be remembered by them, so make them count.
- For openings of informal meetings, pick one:
  - Use a current news event.
  - Localize an experience.
  - Give an interesting anecdote.
  - Use a relevant quotation.
  - Make a dramatic or surprising statement.
  - Pose a rhetorical question.
- For closings, pick one:
  - Sum up key points.
  - Make a prediction.
  - Call to action.
  - Convey your course of action.
  - Stress solutions.
- End on an upbeat note.
- Be yourself and be natural, energetic, passionate, comfortable, and confident.
- Practice, practice, practice.
- Think in headlines.
- Create visual images.
- Be conversational.
- For longer or multi-speaker presentations, provide a visual roadmap to orient the audience at key intervals.
- Acknowledge preconceptions, especially where they contradict with facts and/or experience.
- Have vocal variety, changing up pitch, tone and speed.
- Try lowering the pitch of your voice (speaker anxiety can cause your voice to go higher).

- Enunciate.
- Avoid acronyms and jargon.
- Pause.
- Have a good, erect posture; do not cross your arms.
- Speak to the audience, do not read from the screen.
- Use natural gestures.
- Count items on your fingers.
- Use props and visuals.
- Avoid playing with a pen or other objects.
- Use facial expressions.
- Give examples. Where appropriate, use anecdotes and personal stories.
- Use alliteration. For example:
  - Safety
  - Security
  - Strategy
- Use an anaphora. For example:
  - The best…
  - The best…
  - The best…
- Go for a walk and do breathing exercises before your presentation.
- Fill your lungs, project energy.
- Less is more—keep the presentation short and slides simple.
- Pause before the question and answer session.
- Have a final closing statement ready for the end of the Q&As.
- Visit the room early; check equipment and have backups.
- Have a glass of water ready.
- Relax and have fun.

Appendix G: Laundry List of Question-and-Answer Tips

- Anticipate and practice answering questions in advance.
- Think through and fully research responses in preparation.
- Use a "devil's advocate" or "red team" to prepare. Practice answering the toughest questions you think you might get.
- Pause between giving a presentation and entering the Q&A session.
- Listen carefully.
- Always repeat questions before answering if there is a chance the question was not heard by all.
- Don't repeat offensive phrases or words contained in a question.
- Make your main point first, then expand.
- Keep answers short, but avoid abrupt yes or no answers.
- Do not ramble. Stop when you have made your point.
- Use stories and personal examples.
- You are the expert—speak only about what you know.
- Be positive.
- Be conversational and animated.
- Use understandable language.
- Avoid acronyms and jargon.
- Clarify any unclear issues.
- Demonstrate ownership of your proposal or topic.
- Have prepared points. Know them but don't memorize a statement word-for-word.
- Keep direct, friendly eye contact and a slight smile.
- If you don't know, say so and offer to find out the answer (and then do so!).
- Incorporate teammates in answers where possible (and appropriate).
- Be honest, candid, and truthful. Don't exaggerate.
- Never be evasive or half-truthful.
- Keep your cool under fire.
- Don't argue.

- Have situational awareness. Be up on current events and breaking news important to the audience.
- If you are interrupted before you have finished a response, let the questioner finish and then continue your answer.
- If you get asked several questions at once, pick the one you want to answer and then after answering it, ask the questioner to repeat another of the questions. As an alternative, you can say "Well, you've asked several questions. Let me respond to your main point first..."
- Focus or narrow a question if necessary.
- Do not try to show-up the questioner; be humble.
- Avoid playing with pens or other objects. Remove the temptations beforehand if necessary.
- Remember you are always on record, even during the break or after the session is over.
- Do not be afraid to tell the questioner that he or she is mistaken. If you do not correct an erroneous statement, your lack of response could imply agreement.
- After you answer a question, stop and wait for the next question. Do not be afraid of silence.
- Make you major point even if it wasn't asked for. Answer the question first, and then tag on what you feel needs to be said.
- Watch the audience as you answer a question. Do they look satisfied or confused?
- Be careful about speculating on hypothetical situations.
- Don't repeat the negative part of a question when giving your answer. Stress the positive when possible. After giving your answer, transition to a positive statement.
- Ask for a question to be repeated or restated if it is unclear.
- When someone else on your team is answering a question, pay attention; be interested, look at him or her, not your phone.
- Never appear impatient for the end.
- Listen for guidance masquerading as small talk or polite appreciation.
- Immediately convene a team debriefing at a prearranged site. If not possible, arrange a later call.

*Appendix H: Templates for Responding to Difficult Questions**

Introduction

"Will you guarantee me that_?"
"Isn't it better to be safe than sorry?"
"Why should I trust you?"
"I think it's riskier than you're telling us!"
"Why shouldn't we be frightened about____?"
"You're more concerned about protecting your organization than us!"
"How do we know that someday science will discover something we don't know today?"
"Don't you think you should have told us about this sooner?"
"Promise us that will never happen!"
"You killed my friend!"
"You're lying to us!"
"How would you like it if____?"
"I don't think that's fair!"
"Where do you live?"
"Do you drink the water?"
"You're an idiot!"
"Your policy/plan is wrong!"
"We have a report that contradicts what you just said!"

All of us experience situations where we receive difficult, challenging and sometimes even insulting questions and statements from others. This can occur in your job, your day-to-day chores, your social life, and even family life. Training is necessary to respond to these situations.

You may have said after a challenging conversation with a stakeholder, "I wish I would have known how to respond to that better!"; "If I had only thought of saying that…;" or "There's no way you can respond to that…"

Outlined below are two flexible, hands-on tools to train and prepare for any situation on any issue that will likely include challenging questions and statements. These tools are Generic Categories and a 4-Step Guideline. These tools work hand in hand for any issue and with both internal and external stakeholders.

* U.S. Navy. Undated. Navy and Marine Corps Public Health Center Environmental Programs. *A Risk Communication Primer,* Fulton, K., and S. Martinez, Fulton Communications.

NOTE: The applications of these tools do not apply to media communication. The media communications process is a unique form of stakeholder communication.

Tool #1 – Generic Categories

The Generic Category Tool for Responding to Challenging Questions and Statements is a tool for the best approach on *how to start a response* to questions and statements from any stakeholder on any issue.

All questions and statements fall into one of the 12 categories so it is a tool that can be used throughout a conversation as questions and statements move from category to category. Usually the goal is to have a conversation in Category 11—Factual Questions. However, if the conversation starts or moves to emotional categories such as Categories 1, 3, and 12, this tool provides the best way to start your response and gives you the best chance to eventually have a factual discussion, Category 11, with the stakeholder.

Similarly, challenging questions that usually occur in Categories 3 through 8 are best handled by starting your response as shown in this tool. Again, this gives you the best chance to eventually have a factual discussion.

The table below provides just a few examples of each category, major traps to avoid and *how to start* the response—the last column titled, "Generic Nature of the Response."

The best way to use this tool is to practice with a co-worker familiar with your communications issues.

Category No.	Category Type	Examples	Major Traps	Generic Nature of Response
1	Ventilation—A highly negative emotional state/ anger, irritation, disgust	• "You killed my friend!" • "I have cancer because of you!" • "You don't care about us!"	• Responding too early with factual information • Taking their comments personally • Inadequate nonverbal observation skills to detect if they are calming down	• First, stay with empathy for awhile • Second, if they have calmed down some-what based on your nonverbal observation, use open ended questions • Third, move to facts if they appear to be ready to discuss facts

Category No.	Category Type	Examples	Major Traps	Generic Nature of Response
2	What is the question or statement?	• "Babble, babble, babble." • You cannot figure out what their point or question is	• Assuming you know the question or statement and answering it	• A softball pushback statement such as, "I want to be sure and answer your question, so can you tell me more about…?"
3	Rude but briefly acceptable	• "You're an idiot!" • "Are you a REAL doctor?" • "Where'd you get your birth certificate?" • "You're the agent of Satan!"	• Taking it personally • Not planning ahead of time on what is acceptable and what is not acceptable	• Acknowledge they are upset. "Clearly you are upset. What can I do to help you?" • How long you allow this will depend on several factors; size of the group, percent of people in a crowd being rude NOTE: This category mostly applies to public settings.
4	Negative Allegation That is Not True	• "Why are you lying about...?" • "You're hiding and covering up."	• Pushing back and reinforcing the negative allegation, e.g., "We didn't lie." Or "Why do you think we're lying?" • "We didn't cover up anything!"	• Start with emphasizing the positive reversal, e.g., the opposite of lying is telling the truth, the opposite of covering up is being open/disclosing, etc. "Actually, we told the truth about that."
5	Negative allegation that is true	• "Why did you lie about...?" • "You covered up."	• Defensiveness or denial when in fact the allegation is true • Not getting approval for your response ahead of time from Command, Legal & Public Affairs, and others	• Acknowledge the truth • Emphasize the commitment to corrective action past, present, and future • "We could have done a better job."

(*Continued*)

Category No.	Category Type	Examples	Major Traps	Generic Nature of Response
6	Guarantee/ 100% assurance No risk acceptable	• "Promise me this will never happen again." • Can you guarantee me that...?" • "Why can't you go to zero?" • "Isn't it better to be safe than sorry?"	• Initially saying yes, no or maybe • Saying initially, "There are no guarantees" or "We can't guarantee you that."	• Emphasize your commitment and what you are doing • "What I can guarantee…" • "We're moving towards zero." • "We are making progress on…" • "We learned a lot from that and this is what we changed."
7	Fairness questions	• "Do you think it's fair that I have to drink this water?" • "I don't think it's right that I have to do 'X' because of you."	• Evasive or defensive • Starting with Cost/Benefit discussions • Not always being aware of common ground opportunities	• Be open about your plans, even if the news is bad for them • Be willing to pursue their point if there may be common ground
8	The setup question or statement	• "Where do you live?" • "Have you taken the vaccine?" • "How would you like it if you had to work in this building/ old housing?"	• Trying to avoid the setup point • Not recognizing that their setup is not their underlying issue. It is just a way of them saying, "You aren't in my situation."	• Provide the info they request in the setup and let them go to their underlying issue, e.g., "I live 'X'" or, "You're right, I haven't been in that situation." Or "I don't work in that building."
9	Personal interest that is not relevant (in Group Discussions)	• A question or statement about issue "X" when the discussion/ meeting is about issue "Y"	• Getting into the non-relevant discussion • Abruptly/ rudely changing the subject	• Bridge back to relevant subject followed by possibly expressing a willingness to discuss another time or send to another source • "I'd be glad to discuss that with you another time, but tonight we're here to…"

Category No.	Category Type	Examples	Major Traps	Generic Nature of Response
10	Policy	• "I don't want to..." • "I think I deserve..."	• Talking too much about their situation and possibly misleading them in terms of policy options • Going into details when they may just want a yes or a no	• Stick with a clear statement of the policy and repeat if necessary
11	Factual questions—What? Who? When? Where?	• "When is the next meeting?" • "What are the next steps" • "When will you find out the results of the testing?"	• Jargon	• Provide a simple/direct response • Respond in language understandable to the stakeholder • Know when to stop talking—non-verbal observation skills
12	Fear	• "I'm afraid of…" • "It's really going to get bad." • "I'm not feeling good about this because…"	• Not being truthful about what is not known • Trying to avoid fear	• Tell them what you do know • Be open about what you don't know • Tell them when you'll update them

Tool #2 – 4-step Guideline

The second tool is a 4-Step Guideline. This is a flexible guideline, not a model that you always use in a 4-step linear manner.

1. Empathy
2. Conclusion
3. Facts
4. Future action

Step 1—Empathy

Sometimes it is appropriate to indicate to your stakeholders that you have some idea of what they are saying and/or some sense of their situation. Empathy is not sympathy

Empathy is not sympathy and empathy is not agreement.

and empathy is not agreement. Also, empathy is not "I know how you feel" because you do not know how they feel.

Empathy is your ability to figure out the following: What must their situation be like for them?

To do this, you must "remove yourself" and think about them instead of yourself. Removing yourself means you cannot bring in your personal feelings. Empathy cannot be artificial or fake. It must be genuine. You cannot "pretend" to be empathic to their situation because stakeholders can tell if you are sincere by your non-verbal communication.

Empathetic statements are frequently not necessary. They are most helpful when dealing with anger, fear, crises, distrust, and significant concerns. Empathy statements, if used, should usually be stated before any of the other steps.

Personal connections can be made in an empathy statement only if the connection is 100% relevant. Examples of effective direct connection empathy statements would be, "I live in your neighborhood, too" or "My family also drinks that water" or "I've taken the vaccine" or "My child also attends that school" or "I went out there and saw that." Empathy statements that would not be effective would be, "I work next to your community" or "I would drink that water if I lived here" or "I would take that vaccine."

Major traps in empathy statements are:

- Using personal connections that are not relevant to the listener
- Giving statements that are not genuine—in your words, body language and voice.

Step 2—Conclusion

The conclusion is usually the most difficult step in the 4-Step Response Guideline because in risk communication, the conclusion must be short, simple and precede the facts that support the conclusion. The conclusion should address the underlying point of the question or statement.

The conclusion is usually the most difficult step in the 4-Step Response Guideline.

Examples of good conclusion statements are:

"The water is safe to drink."
"The vaccine is safe and effective."
"I don't know, but I'll find out."
"We've been sharing all the information with you."
"We are doing a lot."
"We don't plan further clean up."
"We could have done better back then."
"You have to take the test."
"The food is safe to eat."
"The policy states that…"
"We cannot provide that to you."
"We can provide that to you."
"You have to wear the respirator."
"One thing that has to happen first is…"
"The cleanup is complete."
"We don't plan to spend any more money."

Also, if you are concerned that they may not be listening to your conclusion, you can use opening phrases such as:

"Our conclusion is…"
"The answer to your question is…"
"What we learned was…"
"The good news is…"
"The unfortunate news is…"
"I'm sorry to say…"

Major traps in the conclusion step are:

The conclusion statement does not address the underlying point or question made by the stakeholder.

The conclusion statement is too long.

Facts are included in the conclusion, e.g., "The water is safe to drink because 'X, Y, Z'." Instead, say, "The water is safe to drink" then pause to see if you were heard. Then, "The reason I say that is (facts)." The facts are delivered separately. First make sure they heard the conclusion.

You can use transition statements between your conclusion and facts:

- "I say that because..."
- "Because we have developed..."
- "The reason for that is..."

Step 3—Facts

Facts support your conclusion. Usually one, two, or three facts are sufficient. There is no right number of facts to support your conclusion. In some instances, you may only have one fact. Other times, you may have several facts and your stakeholders are interested in all of them. That is, they are actively listening. In those instances, use all your facts. It is crucial that you use your nonverbal skills here. As you are speaking, determine whether your stakeholders are listening to you. If not, stop talking about your facts and find out why they aren't listening, "Am I being clear?"

You can use transition statements between your conclusion and facts:

- "The reason I say that is..."
- "Why, because we have developed..."
- "The reason for that is..."

Major traps in this step are:

- Overuse of negative words and phrases unless your purpose is to change behavior.
- The use of what would be considered jargon for the stakeholders.
- Not observing if the stakeholders are listening.

Step 4—Future Action

You may not always have or need a future action in your verbal response. There are many instances where you close/complete the response without a future action. Many times the conclusion is all you need, e.g., *We can't change policy.* However, it is usually important to have a future action when the stakeholders are concerned, fearful, distrustful, worried, or confused.

Future action statements should have a "when," a timing factor. If you don't have a "when," then tell them "when" you'll have a "when." For example, "I'll call you next Friday. I may have that information then." Whatever your future action comment is, it should let the stakeholders know that they will continue to be involved, unless, of course, their point/issue has been resolved.

Good future action statements are:

- "I don't know, but I'll call you tomorrow."
- "I don't know, but I'll let you know at the meeting next Tuesday."
- "I'll be happy to talk to you more after the meeting."
- "There's more information about this on our website/brochure/fact sheet."
- "The next review will be held at 'X' on 'Y' day."
- "We won't know for at least 6 months, but I'll be glad to call/email once a month on our latest outlook."

Major traps in the future action step are:

- Not mention a "when" or "when" you might have a "when."

Summary of the 4-Step Guideline:

- It is a guideline, not a model.
- You may not have conclusion/facts, just a future action. If so, the future action is also your conclusion, "I don't know. I'll call you tomorrow with more information."
- You can use transition statements between steps.
- This guideline is not for media communications.
- The guideline is not effective without good nonverbal skills, self-awareness and observation skills.

References

U.S. Navy. Undated. Navy and Marine Corps Public Health Center. Risk Communication Primer: Tools & Techniques. Appendix H. Two Tools for Responding to Any Difficult Question/Statement on Any Issue from Any Stakeholder in Any Setting. Fulton, K., and S. Martinez, Fulton Communications.

Index

Italicized and **bold** pages refer to figures and tables respectively

K

L

M